The Quality of Urban Air Review Group

Chairman
Prof R M Harrison
The Queen Elizabeth II Birmingham Centenary
Professor of Environmental Health
The University of Birmingham

Prof P Brimblecombe
University of East Anglia

Dr R G Derwent
Meteorological Office

Dr G J Dollard
AEA, NETCEN

Dr S Eggleston
AEA, NETCEN

Prof R S Hamilton
Middlesex University

Mr A J Hickman
Transport Research Laboratory

Dr C Holman
Independent Consultant

Dr D P H Laxen
Independent Consultant

Mr S Moorcroft
TBV Science

Observers

Dr S Coster
Department of the Environment

Ms L Edwards
Department of the Environment

Dr R L Maynard
Department of Health

Secretariat

A R Deacon
Institute of Public and Environmental Health
The University of Birmingham

The authors accept responsibility for the contents of this report but the views are their own and not necessarily those of the organisations to which they belong or the Department of the Environment

ACKNOWLEDGEMENTS

The Review Group is most grateful to the following individuals and organisations for assistance.

John Garland (NETCEN) - main authorship of Chapter 2

David Mark (AEA Technology, Harwell) - main authorship of Chapter 3

Gary Hayman (NETCEN) - main authorship of Chapter 6

Helen Clark, Alison Loader, John Stedman, Ken Stevenson, Paul Willis, Geoff Broughton, Glenn Campbell and David Lee (NETCEN) - data analyses for Chapter 6

Marcus Jones (The University of Birmingham) - data analyses

Alan Turnbull (The University of Birmingham) - information on biological particles

Andrew Clarke (The University of Leeds) - provision and interpretation of air quality data

Chris Miller (Salford University) - air quality trend data

Bob Appleby, Gavin Tringham and Peter Fallon (Birmingham City Council) - provision of air quality data

David Muir (Bristol City Council) - provision of air quality data

Westminster City Council, Planning and Environment Department - provision of air quality data

John Rice and Sean Beevers (South East Institute of Public Health) - provision of air quality data

Michael Grassi and Colin Harris (The Grassi Design Company) - graphic design and illustration work

Executive Summary

Airborne particulate matter is a very diverse material in terms of its physical and chemical properties and there are many sources which contribute to atmospheric concentrations. This report, the Third Report of the Quality of Urban Air Review Group, reviews knowledge of the sources, chemical composition and physical properties and concentrations of airborne particles and examines the implications for control of particulate matter in UK urban air.

Airborne particles are typically very small; they range in size from a few nanometres in diameter to 100 or more micrometres*. Whilst particles from throughout this range can be found in the atmosphere, those capable of remaining airborne for long periods (typically 10 days or so), are in the more limited size range of approximately 0.1-2 micrometres diameter. The size fraction measured by the UK Automatic Urban Network includes particles up to 10 micrometres (termed PM_{10}). The Report reviews the methods available for measuring airborne particulate matter and highlights the performance characteristics of various common instruments.

Airborne suspended particulate matter may be either primary or secondary in its origins. Primary particles are those directly emitted to the atmosphere from sources such as road traffic, coal burning, industry, windblown soil and dust and sea spray. On the other hand, secondary particles are particles formed within the atmosphere by chemical reaction or condensation of gases, and the major contributors are sulphate and nitrate salts formed from the oxidation of sulphur dioxide and nitrogen oxides respectively. Emissions inventories are available for primary particulate matter which indicate that nationally the major sources are road transport (25%), non-combustion processes such as construction, mining and quarrying and industrial processes (24%), industrial combustion plants and processes with combustion (17%), commercial, institutional and residential combustion plants (16%) and public power generation (15%). Industrial sources may thus lead to locally elevated concentrations of PM_{10}. Domestic coal burning in some areas of the UK, for example Belfast, can also lead to elevated levels. Within urban areas generally, however, the influence of road traffic is far more marked and in Greater London it is estimated that 86% of PM_{10} emissions in 1990 arose from this source. The contribution of secondary particles is far more uniform as these are formed relatively slowly in the atmosphere and have a long atmospheric lifetime. Nonetheless, there is a distinct gradient in concentrations with the highest in the south-east of the UK, decreasing progressively to the north and west. Recent data show concentrations of secondary particles some 10% higher in the spring and summer months of the year than the annual average.

For many years measurements of particles in the UK atmosphere have been carried out using the black smoke method. This depends upon the blackness rather than the weight of particles. Originally, this was designed to measure airborne particles derived from the burning of coal, but nowadays in most urban areas the measurements reflect primarily emissions from diesel vehicles. More recently, automatic measurements of PM_{10} have been made and data from the UK Automatic Urban Network show uniformity of annual average concentrations across the major UK urban areas, with mean concentrations within the range of 20-30 micrograms per cubic metre. Exceptional behaviour has been observed in Belfast where coal burning is still a major source of airborne particles, and for a time in Cardiff when construction activity adjacent to the sampling site influenced the measurements. Examination of the highest daily mean concentrations shows a seasonal pattern with the highest values generally occurring in the winter months with variations between urban areas much clearer than for the annual average concentration. There are relatively few data for rural measurement of PM_{10}, but it appears that concentrations are of the order of 10-15 micrograms per cubic metre, reflecting mainly secondary particles.

Receptor modelling techniques based upon an analysis of measurements of airborne particles and their chemical components have been used to identify the major contributors to PM_{10} in urban air. The results indicate that exhaust emissions from road traffic together with secondary particles dominate the $PM_{2.5}$ (small particle) fraction which comprises the major part of elevated PM_{10} concentrations and hence is potentially more important in relation to health impacts. The sources of coarse particles, in the 2.5-10 micrometre range, are less clearly defined, but resuspended street dusts, windblown soils and sea spray particles are major contributors. In the winter months

* 1 nanometre is one millionth of a millimetre; 1 micrometre is one thousandth of a millimetre.

episodes of high PM_{10} are associated almost wholly with raised concentrations of $PM_{2.5}$ arising from vehicle exhaust emissions. In summer, the contribution of secondary particles during photochemical pollution episodes is also important. Road traffic contributes to secondary particle formation through emissions of NO_x. Emissions inventories do not currently give a complete breakdown of the sources of particulate matter in UK urban air as they cannot account for secondary particles, and estimates of emissions from diffuse and natural sources such as dust suspended from road surfaces by wind and traffic-induced turbulence are subject to very great uncertainties.

There are currently many exceedences of the EPAQS recommended limit of 50 micrograms per cubic metre as a running 24 hour average concentration. For example, in Central London between 1992 and 1994 this value was exceeded on 75 days measured from midnight to midnight, or 139 days when measured as a 24-hour running mean in accordance with the EPAQS recommendation. Control of PM_{10} concentrations to within the limit recommended by EPAQS will require major controls on the sources of emission. In winter, our analysis shows that far the major contributor to episodes of high PM_{10} concentration is exhaust emission from road transport. A reduction of about two thirds in emissions from this source would lead to achievement of the air quality standard at background sites in most UK urban areas, but control of roadside concentrations and pollution episodes to meet the EPAQS recommendation would require further emission reductions. For example, to have limited 24-hour average PM_{10} to below 50 µg/m^3 in the December 1991 pollution episode in London is estimated to require a reduction of over 80% in road traffic exhaust emissions. Current projections indicate a cut in emissions of about 52% by 2010. In summer, secondary sulphate and nitrate particles formed from atmospheric oxidation of sulphur and nitrogen oxides appear to be the primary cause of episodes of high PM_{10} concentration, although in summer the contribution of coarse particles, largely resuspended surface dusts, is also important. Numerical model predictions indicate that agreed controls on emission of sulphur and nitrogen oxides across Western Europe will deliver an improvement of about 40% in average concentrations of secondary particles across the UK by the year 2010. This alone is unlikely to be sufficient to bring summer concentrations in line with the EPAQS recommendation.

Our analyses clearly indicate the need for tighter controls on the sources of PM_{10} to achieve the EPAQS standard throughout the UK. In some areas improved controls on industry and a reduction in emissions from the burning of solid fuels for home heating may be required. However, for the majority of UK urban areas the key sources requiring control if air quality is to meet the standards recommended by EPAQS are twofold. The first is road traffic, and our Second Report pointed out the particularly important contribution of diesel emissions in regards of atmospheric particulate matter. The second is that of secondary particles formed in the atmosphere from emissions of sulphur and nitrogen oxides. These are formed from Europe-wide emissions rather than solely those from the UK, and hence international agreements for SO_x/NO_x control beyond those already in place will also be needed.

Table of Contents

1 Introduction 1
    1.1 Sources of Airborne Particulate Matter 2
    1.2 Sizes and Shapes of Airborne Particulate Matter 4
        References 6

2 Properties of Airborne Particulate Material 7
    2.1 Introduction 7
    2.2 Formation of Aerosol Particles 7
    2.3 Properties and Behaviour of Airborne Particles 10
    2.4 Lifetime and Travel of Aerosols 14
        References 14

3 Measurement Methods 17
    3.1 Introduction 17
    3.2 Airborne Particles 17
    3.3 Nuisance Dust 32
        References 35

4 Sources and Emissions of Primary Particulate Matter 37
    4.1 Introduction 37
    4.2 Primary Emission Sources 38
    4.3 Emission Estimates 46
    4.4 Ambient PM_{10} Levels and Emissions 49
    4.5 Future Emission Trends 50
    4.6 Uncertainties 52
    4.7 Biological Particles 52
    4.8 Key Points 53
        References 54

5 Sources and Concentrations of Secondary Particulate Matter 57
    5.1 Introduction to Secondary Particulate Material 57
    5.2 Concentrations of Secondary Particulate Material in the United Kingdom 60
    5.3 Key Points 69
        References 69

6 Concentrations and Trends in Particulate Matter 71
    6.1 Introduction 71
    6.2 Measurements of Particulate Matter 71
    6.3 Measurements and Trends in Particulate Matter 75
    6.4 Correlation Between Different Measures of Particulate Matter 92
    6.5 Behaviour of Particulate Matter Concentrations 95
    6.6 Exceedences of Air Quality Standards and Guidelines 106
    6.7 Spatial Distribution of Particulate Matter 108
    6.8 Ultrafine Particles and Particle Number Counts in Urban Air 115
    6.9 Key Points 116
        References 117

7 The Chemical Composition of Airborne Particles in the UK Urban Atmosphere ...119
    7.1 Introduction 119
    7.2 Composition of Airborne Particles 120
    7.3 Results of Reported Studies 122
    7.4 Data Summary 134
    7.5 Conclusions 134
    7.6 Key Points 135
        Acknowledgement 136
        References 136

8		**Receptor Modelling and Source Apportionment**	**139**
	8.1	Introduction	139
	8.2	Receptor Modelling of Total Suspended Particulate Matter	139
	8.3	Receptor Modelling & Source Apportionment of PM_{10}	142
	8.4	Interpretation of PM_{10} and $PM_{2.5}$ Dataset from North-West Birmingham	149
	8.5	Implications for the Control of Atmospheric Concentrations of PM_{10}	149
	8.6	Summer Data	151
	8.7	Meeting the EPAQS Recommended Standard	151
	8.8	Conclusions	154
	8.9	Key Points	154
		References	155
9		**Effects of Airborne Particulate Material**	**157**
	9.1	Health Effects of Non-biological Particles	157
	9.2	The Effects of Particulate Matter on Visibility	161
	9.3	Nuisance from Soiling	162
	9.4	Effects of Atmospheric Aerosols on Climate Change	164
		References	164
10		**Conclusions**	**167**
		Research Recommendations	**171**
		Glossary of Terms and Abbreviations	**173**
		Terms of Reference and Membership	**175**

1 Introduction

In its First Report (QUARG, 1993a) this Review Group sought to summarise the present state of knowledge with regard to the urban air pollution climate of the United Kingdom. Concern was expressed that a number of pollutants exceed present health-based standards and guidelines. It was also noted that increasingly, as other sources come under better control, the relative importance of motor vehicle traffic as a source of air pollution in our towns and cities has increased. Comparing national and urban inventories of pollutant emissions, there are major divergences attributable to the greater impact of road traffic in urban areas. A tentative look into the future revealed that toughening standards on vehicle emissions will lead to significant improvements in air quality over the next five or ten years, but that unless such emission standards were tightened yet further, an increase in traffic growth could negate the benefits obtained from lower pollutant emissions per vehicle, and that for many pollutants, concentrations might once again start to increase after about 2005.

Having highlighted the important role of road traffic in influencing urban air quality, it was natural that our Second Report (QUARG, 1993b) should deal with a major aspect of this question. The Report reviewed the impact of diesel vehicle emissions on urban air quality and highlighted the fact that with the advent of three-way catalytic converters on new petrol cars, the impact of diesel emissions, at that time mostly from Heavy Duty Vehicles (trucks and buses), on urban air quality would become more marked. A further contribution to this trend was also arising from a substantially increased market penetration of diesel cars which had until 1991 comprised only about 5-6% of the new car market, but increased rapidly to take about 25% of the market at the end of 1993. The report highlighted the very large contribution which diesel emissions make to urban concentrations of particulate matter in general, black smoke in particular, and of nitrogen oxides. Based upon an evaluation of the relative importance of the various vehicle-emitted pollutants and of future trends in emissions, the Review Group expressed concern over the impact of particulate matter and nitrogen oxides from diesels on urban air quality and recommended against an increase in the number of diesel vehicles on our roads unless problems with these two pollutants are effectively addressed.

Since the publication of our Second Report, considerable new information has emerged. A detailed evaluation of the December 1991 pollution episode in London has revealed an increase of about 10% in mortality during the week of the episode compared with the same period in earlier years (Anderson et al, 1995). Statistically significant increases in some illnesses were also established. A careful analysis of the data led to the conclusion that the most probable cause of the adverse health effects was air pollution, and it appears that the deadly smogs of the 1950's and 1960's have not been wholly banished. The 1991 episode appears to be a more modern manifestation of those serious pollution incidents and has in common with them the greatly elevated levels of black smoke particles, but differs in that concentrations of sulphur dioxide were very much lower in 1991, but oxides of nitrogen were probably considerably elevated over concentrations in the 1950's, before routine measurements of nitrogen oxides were made.

The epidemiological evidence for the adverse health effects of airborne particles has also strengthened greatly since the publication of our Second Report. There have been further publications of time series studies in which day-to-day changes in mortality and illness are shown to be related to concentrations of airborne particulate matter. Additionally, two major cross-sectional studies in which mortality rates in different cities have been shown to relate to their air pollution have been published. The Harvard Six Cities Study (Dockery et al, 1993) examined mortality amongst 8,000 known individuals over a period of some 15 years and after controlling for other risk factors such as smoking, excess body weight and socio-economic class, showed a strong relationship between mortality and the concentration of fine particulate matter in the atmosphere. The more recent study based upon the American Cancer Society database of over half a million people used data on fine particulate matter from some 50 urban areas as well as particulate sulphate data from 151 urban areas, showing a clear relationship between mortality and concentrations of the air pollutant (Pope et al, 1995).

As a reaction to the increasing weight of evidence on the adverse health effects of airborne particles, other UK advisory panels and committees have also been active. The Expert Panel on Air Quality Standards has recently concluded its deliberations upon airborne particulate matter and has recommended a standard. In parallel with this exercise, the Department of Health's Committee on the Medical Effects of Air Pollution appointed a Sub-Committee on Particulate Matter which has published a detailed report on non-biological particles and health. These reports provide valuable advice upon the health risks of particulate matter which will form part of the information needed by Government to formulate policy.

The Quality of Urban Air Review Group also has a role to play in providing advice on particulate matter and this, our Third Report, seeks to provide information from our perspective as air pollution scientists. Current research suggests that the health impact of airborne particles is dependent upon their size, but apparently not upon their chemical composition. One hypothesis on the biological mechanism of health injury due to airborne particles implies that the number of particles to which the individual is exposed is more important than their mass (Seaton et al, 1995). In this Report we seek to review knowledge on both the chemistry and physics of airborne particles. We provide information on the measured sizes of particulate matter in the UK atmosphere as well as on the chemical composition, which may yet prove to be significant in the toxicity, but notwithstanding, is a very valuable indicator of the source of the particles.

Any control strategy for airborne particulate matter must be based upon a firm understanding of the origins of those particles and the factors which influence their concentrations in the atmosphere. This report seeks to review the current state of knowledge on these issues. For example, it is clear that airborne particulate matter may be both primary, ie emitted as such into the atmosphere, or secondary, ie formed in the atmosphere from chemical reactions of gaseous precursors such as sulphur and nitrogen oxides. The two kinds of particulate matter show different seasonal patterns and a different dependence upon the control of sources. Here we seek to review the limited knowledge available on the partition between primary and secondary sources, and upon the contributors to the primary particulate matter in the atmosphere.

The review and analysis included in this Report reveals the considerable diversity in the sources of particulate matter and the many factors influencing its atmospheric behaviour. In this context, particulate matter is certainly the most complex of the common air pollutants to understand and may ultimately prove one of the most difficult to control because of its multiplicity of sources. However, by targeting major sources of primary emissions such as road traffic, as well as sources of the sulphur oxide and nitrogen oxide precursors of secondary aerosol, it should be feasible to make progress over the coming few years. Refined and updated projections of future emissions are included in this Report.

1.1 SOURCES OF AIRBORNE PARTICULATE MATTER

At this point in the Report it is appropriate to give a very brief summary of the sources of airborne particles. Those which are relevant will be amplified further in the Report. On a global scale a number of natural sources such as volcanic emission of particles, forest and brush fires and ingress of extra-terrestrial particles into the atmosphere may be significant. In the context of the United Kingdom, and particularly in relation to urban areas, these sources are dwarfed by particles arising from human activity and will not be considered further.

Suspension of Soil Dust

The action of the wind on dry loose soil surfaces leads to particles blowing into the air. This is most obviously seen in association with sand and dust storms common during windy conditions in the world's deserts. These processes do, however, extend to other regions of the globe, although their magnitude is obviously reduced where soils are moist and have vegetation cover. It is estimated that suspension of surface soils causes the introduction of 150 million tonnes of dust per year to the Northern Hemisphere atmosphere and around double this if the Sahara Desert is included. This leads to an atmospheric burden of dust in the Northern Hemisphere, (excluding the Sahara plume) of 3 million tonnes on average. The presence of such particles in the UK atmosphere is manifested by the accumulation of dust on surfaces such as motor cars during dry weather periods. Much enhanced deposition is occasionally seen when the atmosphere carries dust from the Sahara Desert regions. Such particles are generally rather coarse (ie large in size) and frequently have only a limited atmospheric lifetime and range, although their transport from North Africa to the UK shows that this is not always the case.

Factors favouring the suspension of soil dust particles into the atmosphere are an exposed dry surface of fine soil and a strong windspeed. In towns and cities the areas of exposed soil, particularly in town centres, are rather small. However, there are considerable quantities of dusts on road and pavement surfaces which arise from ingress of soil on vehicle tyres and from the atmosphere, the erosion of the road surface itself and degradation of parts of the vehicle, especially the tyres. Because these particles lie on a surface which readily dries, and are subject to atmospheric turbulence induced by passing vehicles, this provides a ready source of particles for resuspension into the atmosphere. The amounts of dust resuspended in this process are extremely difficult to predict or measure as they depend critically upon factors such as the dust loading of the surface, the preceding dry period and the speed of moving traffic. However, the size distribution and chemical composition of particles in the urban atmosphere give a clear indication that this source contributes significantly to the airborne particle loading of our cities.

Seasalt

Breaking waves on the sea cause the ejection of many tiny droplets of seawater into the atmosphere. These droplets dry by evaporation leaving seasalt particles suspended in the air. Whilst these particles are, in the main, rather coarse in size, a minor part of the mass is in particles small enough to have an appreciable atmospheric lifetime, and all parts of the UK are influenced to some extent by particles of seasalt. Clearly, coastal areas are the most affected, but seasalt is measurable even at the most inland of UK locations. In winter months, identification of this source by measurement of the chemical composition of airborne particles is complicated by the use of salt for de-icing the roads. Most road de-icing salt has a chemical composition almost identical to that of seasalt and the two are effectively indistinguishable. Vehicles travelling at high speed on the motorway in wet conditions raise very visible plumes of spray which contain salt if the road has been treated with de-icing salt. Although the process is less visible, it occurs also on urban roads at lower traffic speeds.

Anthropogenic Primary Particles

All combustion and metallurgical processes and many other industrial operations lead to the emission of particles into the atmosphere. If these particles are directly emitted from a source, they are termed primary. The largest individual contributor of particles is combustion processes, although the mode of formation and chemical composition of the particles varies greatly from one source to another. Thus, in the case of coal combustion, the major part of the particle emissions arises from so-called fly-ash, which is fine particles of mineral material contained in the coal. In the case of emissions from diesel vehicles, the particles comprise largely elemental and organic carbon with some sulphate and water. Fuel is introduced into the combustion chamber of the engine as a spray of fine droplets, each of which leaves a tiny residual particle of unburned and pyrolysed fuel and oxidised sulphur which is emitted from the exhaust. Combustion of other more volatile carbon-containing fuels can lead to formation of carbon-based particles through combination of carbon atoms within the combustion zone. High temperature metallurgical processes and refuse incineration cause the formation of fine metal-rich particles by the condensation of cooling vapours. Despite the best efforts of arrestment plants, a small proportion of such particles inevitably enter the atmosphere. The emissions of anthropogenic primary particles are those most easily quantified and inventories of emission are available which will be elaborated upon later in this report.

Many other industries and processes also lead to emissions of primary particles. These include, for example, mining, quarrying, construction and demolition.

Anthropogenic Secondary Particles

The term secondary particles is used to describe those particles that are formed within the atmosphere, mostly from the chemical oxidation of atmospheric gases. The most prevalent secondary particles are sulphates formed from the oxidation of sulphur dioxide. The first product of this oxidation process is a mist of sulphuric acid droplets, although there is generally ample ammonia available in the UK atmosphere to neutralise this first to ammonium bisulphate and subsequently to ammonium sulphate. This compound is one of the most abundant substances in UK air.

Nitrogen dioxide may also be oxidised in the atmosphere to form nitric acid. The latter is more volatile than sulphuric acid and exists in the atmosphere in the gas phase, but when reacted with ammonia can form particles of ammonium nitrate. It can also react with seasalt particles to form sodium nitrate. Another less abundant form of secondary particles comprises ammonium chloride formed from the reaction of ammonia with hydrochloric acid gas emitted from combustion of coal and municipal incineration.

Chemical reactions of hydrocarbons within the atmosphere can also lead to the production of involatile or semi-volatile products which contribute to the loading of atmospheric particles. The composition of such particles and the chemistry leading to their formation has been studied in the context of pollution in Southern California, but there has been no research on the matter in the UK, and whilst it is likely that secondary organic materials are a component of particles in the UK atmosphere, there is currently no data upon the concentrations or composition.

Biological Particles

It has long been known that the atmosphere acts as a medium for transport of a variety of biological particles. The sizes of such particles cover a very wide range. Vegetation canopies are a major source of natural biological particles such as those derived from fungal, bacterial and viral plant pathogens as well as pollen grains derived from the flowers of wind-pollinated plants. There are well established networks for sampling, counting and identifying airborne pollens and spores in the UK atmosphere. Information on bacteria and viruses is much more scarce, generally being related to investigations of specific problems. However, the recent development of DNA-based techniques in molecular biology offers the opportunity for more extensive work in this field.

Airborne biological particles can cause disease. The effects produced are dependent on the nature of the particles: viral, bacterial, fungal or pollen. While such effects are of important health significance they are not a facet of urban air pollution specifically and are outside the remit of this Report.

1.2 SIZES AND SHAPES OF AIRBORNE PARTICLES

Airborne particles cover a very wide size range from a few nanometres (billionths of a metre) to tens of micrometres (millionths of a metre). The recently published EPAQS recommendation relates to PM_{10}, or particles passing an inlet of defined characteristics with a 50% sampling efficiency at 10 micrometres (μm) aerodynamic diameter. Thus, to a good approximation, PM_{10} is particulate matter smaller than 10 μm diameter, and usually comprises the majority of airborne particle mass. Particles outside of this size range are relatively very large in size and have little impact on health. Within the PM_{10} size range, particles of less than 2.5 μm aerodynamic diameter (known as $PM_{2.5}$) are normally described as fine, whilst the 2.5 to 10 μm fraction is termed coarse. The fine particles are capable of reaching the deepest part of the lung, whilst coarse particles generally deposit in the upper airways. As a rough generalisation, particles generated from combustion and condensation of vapours are mostly in the fine fraction, whilst particles from mechanical break-up of solids and liquids are coarse.

The physical form of PM_{10} may be assessed by electron microscopy. Some examples appear in

Figure 1.1 Airborne Particles Collected on the Surface of a Nuclepore Filter from the Atmosphere of Birmingham. The Dark Circles are Pores in the Filter, about 0.4 micrometres in Diameter.

(a) Mixed small particles. The branched chain structures are probably vehicle-generated particles.

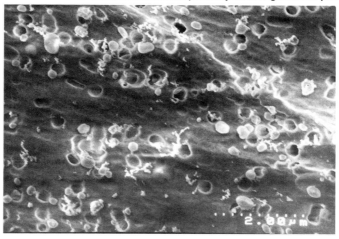

(b) Close-up of a cluster formed from coagulation of many small particles, each about 50 nanometres in diameter.

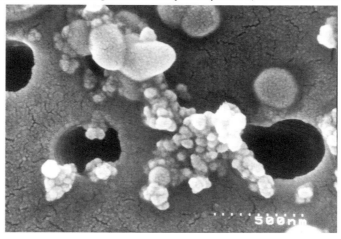

(c) Two very large particles, about 5 micrometres in diameter. One (on the left) appears to be a layered mineral, whilst the other is a massive cluster of tiny particles.

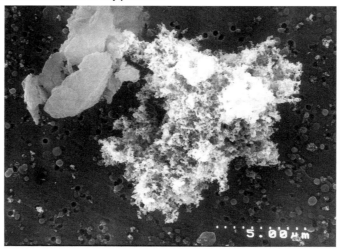

Figure 1.1. This exemplifies the wide variety of sizes and shapes of particles typically encountered in the atmosphere.

REFERENCES

Anderson HR, Limb ES, Bland JM, Ponce de Leon A, Strachan DP and Bower JS (1995) **Health Effects of an Air Pollution Episode in London, December 1991**, Thorax, **50**, 1188-1193.

Dockery DW, Pope CA, Xu, X et al (1993) **An Association Between Air Pollution and Mortality in Six US Cities**, N Engl J Med, 329, 1753-1759.

QUARG (1993a) **Urban Air Quality in the United Kingdom**, First Report of the Quality of Urban Air Review Group, QUARG, London.

QUARG (1993b) **Diesel Vehicle Emissions and Urban Air Quality**, Second Report of the Quality of Urban Air Review Group, QUARG, Birmingham.

Pope CA, Thun MJ, Namboodiri MM, Dockery DW, Evans JS, Speizer FE and Heath CW (1995) **Particulate Air Pollution as a Predictor of Mortality in a Prospective Study of U.S. Adults**, Am J Respir Crit Care Med, 151, 669-674.

Seaton A, MacNee W, Donaldson D and Godden D (1995) **Particulate Air Pollution and Acute Health Effects**, Lancet, 345, 176-178.

2 Properties of Airborne Particulate Material

2.1 INTRODUCTION

Simple physical processes and properties profoundly influence the behaviour of atmospheric particles, and their effects on our health and environment. The air near the ground contains a variable mixture of diverse particles, some solid and some partly or wholly liquid, including many of the pollutants of current concern. Although the properties of each particle differ according to its size, shape and composition, the effect of particle size is strongly dominant.

The purpose of this Chapter is to provide a brief overview of the properties of aerosol particles that are of major importance. The persistence in the air affects the atmospheric concentration and the distance travelled before deposition. If this distance is sufficiently great, the effects of pollution are international or even intercontinental issues. The same characteristics determine how far particles follow the flow of inhaled air into the respiratory system, and whether the particles are retained in the nose, bronchi or lungs, or are simply exhaled harmlessly. Light absorption and scattering by particles influence the visibility of distant objects in conditions of haze, the colour of the sky (when cloud is absent) and the appearance of the landscape. The bibliography includes a few references that provide more detail of the general properties and effects of aerosols.

2.2 FORMATION OF AEROSOL PARTICLES

2.2.1 Size Scale

The mechanism of formation is crucial in influencing the size of aerosol particles. Several distinct mechanisms contribute particles to the atmosphere, and as a result the atmospheric aerosol spans a very wide range of sizes. The mechanisms operate differently for various materials and as a result the composition varies markedly over the size range.

The units used to describe aerosol particles, and the range of sizes involved, are illustrated in Figure 2.1. The range of sizes encountered is limited at the lower end by the size of a cluster of half a dozen or so molecules: this is the smallest entity of the condensed phase that can exist. The upper limit of size is approximately the size of a sand grain. Such particles fall quickly to the ground and normally travel only short distances while airborne. They cannot be properly described as suspended in the air except in extreme, high wind conditions.

2.2.2 Condensation and Nucleation

The finest particles result from the condensation of small numbers of involatile molecules. They may be formed in combustion processes and in evaporation from hot surfaces. Smokes and fumes from metallurgical processes, coal and oil burning, motor engines and even electric fires and cookers, contain numerous particles in this size range. Such particles also may result from gas-phase reactions (such as oxidation of SO_2 to form H_2SO_4) in the free atmosphere that form involatile molecules.

When molecules of an involatile substance are present in concentrations that far exceed the saturation vapour pressure, they have a strong tendency to condense. If a molecule encounters other molecules of the same substance it may combine to form a condensation nucleus - a very small particle. The molecule may meet a nucleus that already exists, or a particle of another substance, and condense causing growth in size. In practice nucleation is self-limiting, since once a sufficient number of nuclei have formed, further condensation is likely to occur on existing nuclei rather than form new ones.

These very fine particles (of order 1 to 10 nm diameter) are very light (see Figure 2.1). Where measurable concentrations of vapour condense in smoke or fume, they are generated in vast numbers - 10^5 to 10^6 per cubic centimetre of air is common in urban and industrial regions. They collide as a result of their rapid Brownian motion, adhere and agglomerate to form larger droplets, or the long dendritic chains of particles commonly seen in smoke. Further involatile matter, such as the sulphuric acid formed in sulphur dioxide oxidation, may condense on these particles. These processes gradually result in particles in the 'accumulation mode', roughly 0.05 to 2 µm diameter.

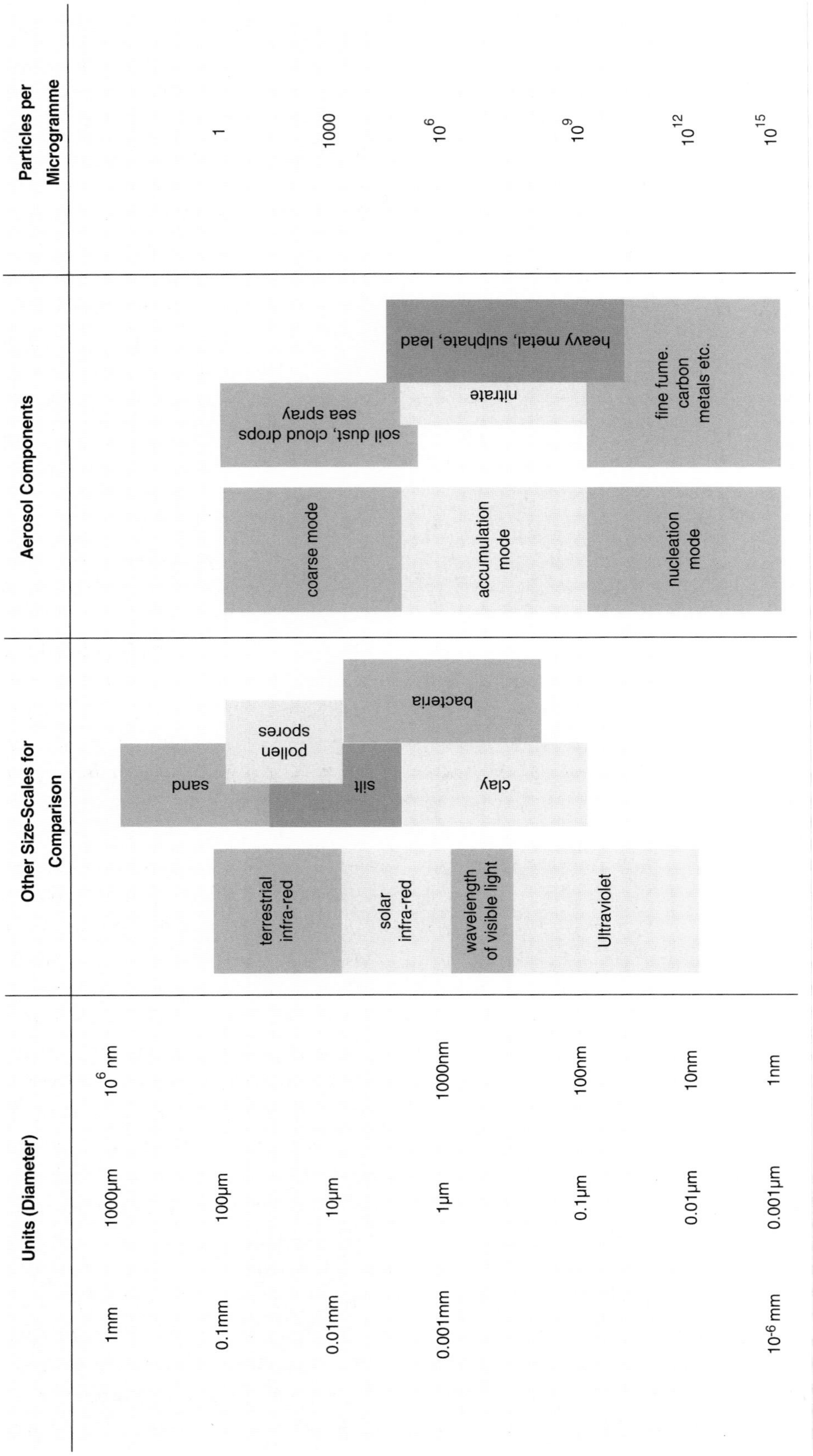

Figure 2.1 Aerosol Size-Scales.

Units (Diameter)		Other Size-Scales for Comparison	Aerosol Components	Particles per Microgramme
1mm				
1000μm	10⁶ nm			
0.1mm				
100μm				1
0.01mm				
10μm		sand	coarse mode / soil dust, cloud drops, sea spray	1000
0.001mm		pollen spores / silt / clay		
1μm	1000nm	terrestrial infra-red		10^6
		solar infra-red / bacteria	accumulation mode / nitrate	
	100nm	wavelength of visible light		10^9
0.1μm		Ultraviolet	heavy metal, sulphate, lead	
0.01μm	10nm			10^{12}
10^{-6} mm			nucleation mode / fine fume, carbon, metals etc.	
0.001μm	1nm			10^{15}

2.2.3 Comminution

A number of processes reduce massive material to fine particles: the energy requirement depends on the surface area of the particles produced and the large amount of energy that would be required to generate great numbers of very small particles is usually prohibitive. In practice particles smaller than about 5 µm diameter are rarely generated this way, although evaporation of spray drops may result in particles a few times smaller.

In the natural world, various mechanisms result in the fragmentation of rocks, the dispersal of soil in the air or the formation of sea spray. Often wind energy is the agent involved in the final step of rendering particles airborne, and often in the fracturing of bulk material.

Man intensifies this process locally by various activities including quarrying, crushing and grinding of ores and other substances, raising dust in agriculture and road transport, spraying etc. In general, the result of artificial or natural dispersal of dust and spray in the atmosphere is a peak in the size distribution in the aerosol in the region of 10 µm diameter.

2.2.4 Fly Ash

This is a significant source of aerosol in some areas and is worth a special mention. Strictly, fly ash is a special case of a comminution aerosol. The fuel is reduced to small dimensions by crushing (coal) or spraying (oil) and suspended in the flame. Carbon and organic matter is lost by oxidation to leave involatile components as particles, which often show evidence of melting. This aerosol component is often accompanied by a fine condensation aerosol.

2.2.5 Cloud and Cloud Chemistry

Although rain and other forms of precipitation result in removal of a large fraction of the aerosol particles from the atmosphere, many clouds do not produce rain, and precipitation is frequently seen to evaporate before reaching the ground. Chemical processes within cloud droplets produce additional aerosol material. Cloud droplets condense onto pre-existing aerosol particles, and the result of droplet growth, cloud chemistry and evaporation is therefore expected to increase the size of the particle and perhaps to change its composition. Similarly, in cloud there are increased opportunities for aggregation of aerosol particles, with further growth in size. There is also some evidence that nucleation (probably of sulphuric acid solution droplets) is favoured at the high relative humidities within or close to clouds, where new, small particles are formed. The effects of clouds on the size distribution have not been quantified, but it is likely that particles chiefly in the accumulation mode increase in size as a result of cloud processes.

2.2.6 Size Distribution

The atmospheric aerosol shows evidence of the several categories of sources in its size distribution. Typically, three major components are recognisable:

The first (nucleation mode) is attributable to the nucleation process described above. This may contain very large numbers of particles of ~10 nm diameter, but because of the small size of each particle the mass in this component is often a small fraction of the total aerosol mass concentration.

The second component is often called the accumulation mode. Particles roughly in the size range 0.05 to 2 µm diameter are long-lived in the atmosphere since the removal mechanisms are least efficient in this region (see section 2.3 below). Coagulation and condensation leads to growth of particles in the nucleation mode into this region, and often a significant fraction of the aerosol mass accumulates in this size region. Particles in this size range are important vectors for long range transport (because they are long-lived) and are efficient light-scatterers, so they are often dominant in optical effects such as visibility.

Finally, the coarsest particle peak in the atmosphere comprises particles of ~10 µm extending to about 100 µm in diameter. These are shorter-lived, very variable according to local conditions, and are likely to travel distances typically of metres to hundreds of kilometres according to size and wind speed. They may contribute substantially to aerosol mass, although the number of such particles is often small.

These components can overlap to comprise the broad size distribution observed in the atmosphere (eg Figure 2.2). The greatest number of particles is almost invariably in the region smaller than 0.1 μm diameter. The mass of particles is predominantly in the accumulation and coarse particle regions, which may make comparable contributions to the total. Even the most remote, unpolluted parts of the atmosphere contain particles at total concentrations of order 300 per cm³ of air, while urban air may have number concentrations about 1000 times larger.

2.3 PROPERTIES AND BEHAVIOUR OF AIRBORNE PARTICLES

2.3.1 Fundamental Principles

The physical properties of small particles, suspended in air, provide the scientific background for understanding the behaviour of the atmospheric aerosol. In this section some of the relevant physical properties are introduced.

(i) **Brownian Motion**: Small particles suspended in still air are seen to move about erratically. This motion results from the variations in the number and directions of impacts of air molecules on the surface of the particle. The smallest particles are most responsive to the tiny impulses from individual molecular impacts, and the effect decreases with particle size (see Figure 2.3).

The random motion that results causes particles to disperse slowly in still air, and to collide with each other and with surfaces: they adhere when they touch, and these contacts result in coagulation to form larger particles, and deposition onto surfaces.

(ii) **Sedimentation**: Solid and liquid particles are about 1000 times denser than air at ground level. They cannot be buoyed up by the air and must fall through it, whatever their size. However, the gravitational settling of the particles is resisted by the viscosity of air, and each particle falls at a constant rate determined by its shape, size and density. The turbulent fluctuations in wind speed near the ground include upward motions but these are rarely much faster than 50 cm/s. Particles that fall through the air

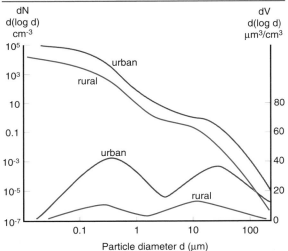

Figure 2.2 Typical Size Distributions of Atmospheric Aerosol in Urban and Rural Areas.

Source: based on Jaenicke, 1993.
Note: The upper curves indicate number distributions, on a logarithmic scale (left hand scale). The lower curves indicate volume distributions (right hand, linear scale), roughly equivalent to mass distribution in units of μg/m³.

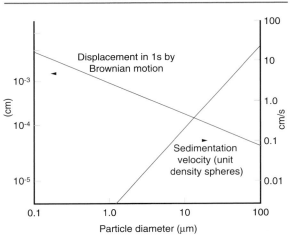

Figure 2.3 Sedimentation Velocity and Brownian Motion, as a Function of Particle Diameter.

as fast as this cannot be lifted far away from the ground surface, and if raised by some disturbance must soon fall out.

(iii) **Impaction**: Close to the ground the wind must flow around obstacles, such as the leaves and stems of trees and grass, irregularities in the soil surface, as well as buildings and other man-made objects. The inertia of the larger particles present in the air prevents them from following the flow around some obstructions to flow. The particles impact against obstructions and may adhere or (usually less often) bounce. The efficiency of this process increases with the air velocity, and particle size, and decreases with the size of the obstacle.

This process is significant in contributions to the deposition of particles onto surfaces, but also is used in sampling particles for measurement. The commonest size-selective samplers are cascade impactors, in which the largest particles are collected first by impaction from a jet of air, usually against a flat plate. In subsequent stages of the instrument, progressively faster air jets separate progressively smaller particles onto impaction plates. The aerosol is size-segregated according to its inertial characteristics. Strictly, the separation depends also on density and particle shape, and the result is specified in terms of the 'aerodynamic diameter' which is the size of a spherical particle of the same density as water with the same inertia.

(iv) **Interception**: Small particles following the airflow around an obstacle may make grazing contact with the surface just because the flow brings them within touching distance. This is an inefficient process, partly because viscosity retards the flow so close to the surface. However it is enhanced by the presence of fine surface structure (leaf hairs, surface roughness of soil, building materials etc) and is important for particles for which Brownian motion and impaction are inefficient.

(v) **Optical Properties**: Impairment of the visual quality of the atmosphere due to haze is often the most immediately apparent effect of air pollution. Particles may both absorb and scatter light, and both affect the visibility of distant objects and the appearance of the sky and landscape. Only particles containing elemental carbon (soot) absorb light significantly, and scattering is by far the dominant process. Even the molecules of clean air scatter light to some degree, explaining the blue colour of the sky in clear, unpolluted conditions. Molecular scattering (and scattering by the smallest particles) is much more effective for blue light than red, being proportional to the fourth power of the ratio (radius/wavelength). For larger particles, the dependence on radius is more complex. When the radius becomes similar to the wavelength, interference between light, refracted through various parts of the particle and diffracted around its edges, causes peaks and troughs in the total scattering, and complex patterns in the distribution of scattered light. Particles a few times larger than the wavelength scatter light equivalent to twice their cross sectional area. When particle mass is taken into account, the most efficient scattering is achieved by particles close in size to the wavelength (Figure 2.4). Scattering by such particles is not much affected by wavelength - hence the whitish appearance of dense haze.

Figure 2.4 Mass Extinction Coefficient, as a Function of Particle Diameter.

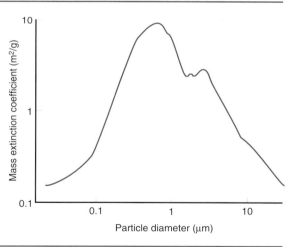

Source: Based on Harshvardhan, 1993, Ch 3 of 'Aerosol-Cloud-Climate Interactions, (PV Hobbs, ed) Academic Press, London and New York.
Note: The mass extinction coefficient (the example shown is for slightly absorbing spherical particles of refractive index 1.53-0.01i) is a measure of the effectiveness of particles in obscuring distant objects, or reducing the brightness of the direct beam from a distant light. For a given mass concentration, particles of about 0.3μm diameter have the most effect on light obscuration, and particles smaller than 0.1μm or larger than 1.0 μm are many times less effective.

(vi) **Uptake of Water Vapour**: Many aerosol particles consist largely of soluble matter. Sulphuric acid or ammonium sulphate and ammonium nitrate comprise a large fraction of the accumulation mode in industrialised regions, and sea salt is an important constituent in coastal regions. These hygroscopic substances take up water vapour at moderate and high relative humidities (Fig. 2.5). This results in an important increase in particle diameter, and influences many processes (light scattering, deposition) that depend on particle size.

In fog and cloud, the air cools until slightly supersaturated with water vapour. Each aerosol particle has a critical supersaturation: up to this point, the particle size increases with relative humidity (supersaturation is relative humidity minus 100%) maintaining equilibrium between the hygroscopic properties of its constituents and the effect of curvature of its surface, which favours evaporation. However, once the critical supersaturation is exceeded the particle continues to take up water vapour, limited only by availability of vapour. The particle grows to form a cloud droplet several microns in diameter. In practice, only a fraction of the larger aerosol particles grow to form cloud or fog droplets. They exhaust the supply of vapour before smaller particles reach their critical supersaturations. The effective particles are called cloud condensation nuclei, and are generally from the accumulation size range. Polluted air with large numbers of nuclei results in fogs or clouds with large numbers of small droplets; speculatively, these have greater persistence and opacity, and reduced likelihood of rain.

2.3.2 Deposition

A combination of processes removes particles from the air to the surface of the earth. Rain collects particles from the atmosphere, as can be demonstrated by analysis of samples of rainwater. Independently of rain, snow and other forms of precipitation, particles are brought to earth by gravity and by several other processes, collectively called dry deposition. Wet deposition, dependent on precipitation, is episodic, since in the UK precipitation occurs less than 10 per cent of the time, but removes particles from a great depth of the atmosphere. Dry deposition operates continuously but only at the surface. Together these categories of deposition must account for removal of all particles in the atmosphere.

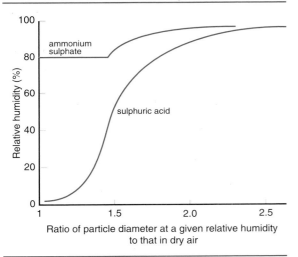

Figure 2.5a Illustration of the Growth of Hygroscopic Particles with Increasing Relative Humidity.

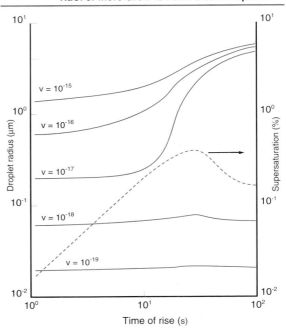

Figure 2.5b. The Growth of Cloud Droplets when an Aerosol of Sodium Chloride Particles of Different Sizes is Carried into a Cloud in a Steady Updraft. In this Example only Particles of 10^{-17} moles of NaCl or more Grow to Form Cloud Droplets.

Source: Adapted from BJ Mason, 1971, The Physics of Clouds, Clarendon Press, Oxford.

2.3.3 Dry Deposition

Near the surface of the earth, the air is mixed rapidly by turbulent motions, generated by the friction of the wind at the surface. The buoyancy of warm air, heated by contact with the surface in daytime adds to the turbulence, while, on clear nights cooling of the surface leads to an adjacent cold layer of air that inhibits turbulence.

The mixing distributes particles through the boundary layer, which usually extends to a height of about 1 km, and is the air layer that is directly affected by friction and heat exchange at the earth's surface. Away from the effects of sources the concentration in this layer would be roughly uniform with height, and in particular, turbulence maintains the concentration near the surface. However, immediately adjacent to a surface, air is retarded by viscosity and must flow parallel to the surface. Particles are carried though this layer, a fraction of a mm thick, inter alia by gravity, Brownian motion and impaction, or contact the surface by interception.

Once particles touch the surface, they are retained by surface forces (although large particles may bounce, they are likely to be trapped following multiple bounces). The combination of mechanisms and the resulting rate of deposition is illustrated in Figure 2.6

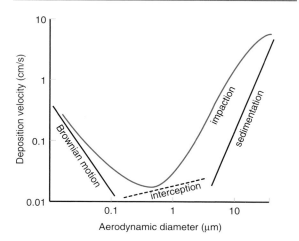

Figure 2.6   Dry Deposition Velocity to Grassland as a Function of Particle Size.

Note: The aerodynamic diameter indicates the inertial and sedimentation behaviour of particles, taking density and shape into account. The diagram shows how different mechanisms combine to remove small and large particles more effectively than those of intermediate size.

The deposition rate is generally expressed as the deposition velocity

$$V_d \text{ (m/s)} = \frac{\text{Flux Density to the Surface (g/m}^2\text{/s)}}{\text{Concentration at Reference Height (g/m}^3\text{)}}$$

and the surface includes any vegetation, buildings or other features standing on a m² of land.

The deposition velocity depends on surface geometry, wind speed and atmospheric stability (the last is controlled by the heating of the surface by the sun, or its cooling at night) but the shape of the curve is generally much the same. The conspicuous minimum at about 0.05 to 2.0 μm diameter coincides with the accumulation mode size range.

2.3.4   Wet Deposition

Falling raindrops collect particles from the air that they traverse by much the same mechanisms. In particular Brownian motion, interception and impaction cause rain to scavenge particulate contamination from the air. As in dry deposition, the capture rate is minimum for particles roughly of 0.05 to 2 μm.

However, wet deposition is more difficult to investigate than dry deposition and our knowledge of wet deposition mechanisms is less complete. Processes within clouds differ from those below cloud in important respects. Rain clouds form in air that has risen from the surface, carrying moisture and aerosol from various sources. The droplets that constitute cloud form by condensation of water vapour on particles of roughly accumulation mode size, resulting in an order of magnitude increase in particle diameter. Generally in the UK, rain formation involves formation of ice crystals, larger than the more numerous water droplets, in the higher parts of the cloud. The falling ice crystals (or snow flakes) collect cloud droplets by impaction and interception before melting to form rain drops. Thus the particles involved as condensation nuclei have a high probability of being scavenged. Together with below-cloud processes, in-cloud scavenging ensures that all parts of the aerosol size distribution are subject to substantial removal rates during rain. Analysis of rain confirms that particulate tracers of all sizes are removed. Averaging wet and dry days together, the results imply an effective removal rate in rain over the UK of about 7% per day for particles of about 1 µm diameter and 15 to 30% per day for particles of 3 to 5 µm diameter and larger.

2.4 LIFETIME AND TRAVEL OF AEROSOLS

From the above discussion, it is possible to indicate the length of time that aerosol particles are likely to spend in the atmosphere (Figure 2.7). The smallest particles (1 nm diameter) last only for some 10 minutes, but their loss is due to agglomeration with other particles (and growth into the accumulation size range) and not loss of material from the atmosphere. In the accumulation size range, particles are likely to be removed from the lower atmosphere by rain in about 10 days (dry deposition alone would take 100 to 1000 days).

Larger particles (say 10 µm diameter) are likely to be airborne for 10 or 20 hours before dry deposition removes them. In the lower troposphere the mean wind speed is about 7 m/s, so the larger particles travel distances of 20 or 30 km while the smaller particles (0.1 to 1 µm) may travel several thousand km.

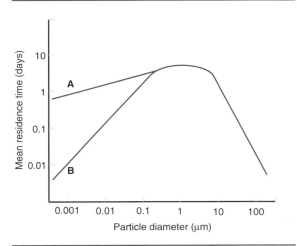

Figure 2.7 The Residence Time of Atmospheric Particles Entering the Boundary Layer.

Source: Adapted from Jaenicke, 1993.
Note: Curve A represents the deposition of aerosol particles to the ground, while curve B also includes the effect of coagulation, which reduces the number of small particles without removing their substance from the air.

Atmospheric conditions are very variable, and, while these values are useful to provide an indication of the average consequences of the emission of pollution to the atmosphere, they should be applied with caution to any particular event or trace substance.

A proportion of particles escape from the boundary layer to higher levels in the troposphere, often as a result of involvement in convective clouds or uplift of air in frontal systems. Those that are not removed in rain are likely to have longer residence times, due to the remoteness of the surface and the lower frequency of falling rain. The small fraction that reach the tropopause (about 8 to 15 km high) may remain airborne for times approaching a year.

The combined effect of persistence and transport of atmospheric aerosols was vividly illustrated by ^{137}Cs from the Chernobyl accident: the primary particles were measurable for over 2 months, showed a mean residence time of about 10 days, and reached all parts of the northern hemisphere.

REFERENCES

Finlayson-Pitts BJ, and Pitts JN (1986) **Atmospheric Chemistry: Fundamentals and Experimental Techniques, Chapter 12: Particulate Matter in the Atmosphere: Primary and Secondary Particles**, John Wiley and Sons, New York.

Jaenicke R (1988) **Aerosol Physics and Chemistry, in 'Numerical Data and Functional Relationships in Science and Technology', Landolt Bernstein New Series V: Geophysics and Space Research, 4: Meteorology (G Fischer, ed.); Physical and Chemical Properties of the Air**, 391-457, Springer, Berlin.

Jaenicke R (1993) **Tropospheric Aerosols, Ch. 3 of 'Aerosol-Cloud-Climate Interactions'**, (PV Hobbs, ed) Academic Press, London and New York.

Junge CE (1963) **Air Chemistry and Radioactivity**, Academic Press, London and New York.

Preining O (1993) **Global Climate Change Due to Aerosols, Ch. 3 of 'Global Atmospheric Chemical**

Change'(CN Hewitt and WT Sturges, eds), Elsevier Applied Science, London and New York.

Twomey S (1977) **Atmospheric Aerosols**, Elsevier Scientific Publishing Company, Amsterdam, Oxford and New York.

3 Measurement Methods

3.1 INTRODUCTION

There is a wide range of samplers available for monitoring particles. The choice though depends crucially upon the purpose of the monitoring. There are two basic reasons for particle monitoring that are dealt with in this Chapter:

- human health effects

- nuisance effects

In the case of human health it is the airborne concentration of particles that needs to be measured, while for nuisance effects it is principally deposited particles that have to be determined. The methods considered are designed to measure the quantity of material in the air or the amount deposited. Methods for chemical or physical analysis of collected particles are not covered.

The potential for particles to cause health effects is related to their size. Particles up to 100 μm enter the body during breathing, but it is only the very small particles, below about 5 μm aerodynamic diameter that can reach deep into the lung. It is widely accepted that it is these very small particles that have the main potential for causing health effects. It is therefore very important to define the size of the particles that are to be measured. It is no longer sufficient to measure what is termed "total suspended particulate" (TSP) or "suspended particulate matter" (SPM), as the size fractions being sampled are not specific but depend upon the equipment used to collect the sample. The current focus of health-related sampling of particulate matter is on PM_{10}, and this will therefore be the focus of the measurement methods described in this Chapter. There is also a growing interest in the finer fraction of particles, such as $PM_{2.5}$, so some consideration will be given to sampling of this fraction.

3.2 AIRBORNE PARTICLES

3.2.1 Scientific Framework

For health effects that are suspected to have arisen from particles entering the body through the nose and mouth during breathing, one must use a sampler whose performance mimics the efficiency with which particles enter the nose and mouth and penetrate to the region in the body where the harmful effect occurs. Workers in the occupational hygiene field have realised this for some years and defined the respirable fraction for those particles that penetrate to the alveolar region of the lung and cause diseases such as pneumoconiosis, silicosis, asbestosis, etc.

Since the early 1980s an ad-hoc working group of the International Standards Organisation has been formulating health-related sampling conventions for airborne dusts both in the ambient atmosphere and in the workplace. The final agreed conventions have passed through all stages of the approval procedure and should soon become available as International Standard IS 7708 (ISO, 1994). They are defined in Figure 3.1 and comprise four main fractions:

Inhalable Fraction (E_I) is defined as the mass fraction of total airborne particles which is inhaled through the nose and/or mouth. It was derived from wind tunnel measurements of the sampling efficiency of full-size tailor's mannequins and replaces the very-loosely defined "total" aerosol fraction used previously. For ambient atmospheres it is given by:

$$E_I = 0.5 (1 + \exp [-0.06D]) + 10^{-5} U^{2.75} \exp (0.05D)$$

where D is the aerodynamic diameter of the particle (defined as the diameter of an equivalent spherical particle of density 10^3 kg/m³, which has the same falling speed as the particle in question), and U is the windspeed (up to 10 m/s).

Thoracic Fraction is defined as the mass fraction of inhaled particles penetrating the respiratory system beyond the larynx. As a function of total airborne particles, it is given by a cumulative lognormal curve, with a median aerodynamic diameter of 10 μm and geometric standard deviation of 1.5.

Respirable Fraction is defined as the mass fraction of inhaled particles which penetrates to the unciliated airways of the lung (alveolar region). As a function of total airborne particles, it is given by a cumulative

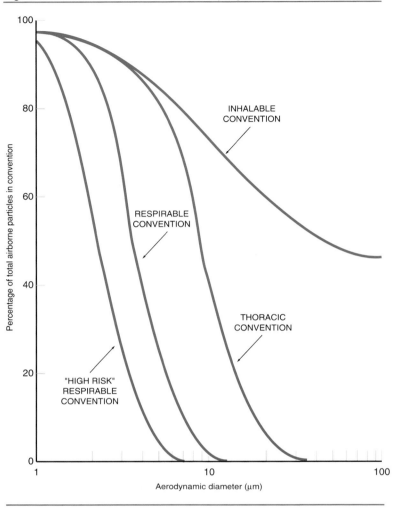

Figure 3.1 ISO Health-Related Particle Sampling Conventions IS 7708 (1994).

lognormal curve with a median aerodynamic diameter of 4 µm and a geometric standard deviation of 1.5.

'High Risk' Respirable Fraction is a definition of the respirable fraction for the sick and infirm, or children. As a function of total airborne particles, it is given by a cumulative lognormal curve with a median aerodynamic diameter of 2.5 µm and a geometric standard deviation of 1.5.

These conventions provide target specifications for the design of health-related sampling instruments, and give a scientific framework for the measurement of airborne dust for correlation with health effects. For example, the inhalable fraction applies to all particles that can enter the body, and is specifically of relevance to those coarser toxic particles that deposit and dissolve in the mouth and nose. The respirable fraction, on the other hand, relates to those diseases of the deep lung, such as the pneumoconioses, whilst the thoracic fraction may be relevant to incidences of bronchitis, asthma and upper airways diseases. Until recently, this philosophy has not been taken on board by the environmental community for health-related sampling.

In the European Union (and thus the UK), ambient airborne particle sampling has been guided by the Sulphur Dioxide and Suspended Particulate Directive (80/779/EEC). This Directive makes no reference to any particular health-related aerosol fraction. The measurement methods accepted are either for 'black smoke' or 'suspended particulates' by a gravimetric method. The EC Ambient Air Quality Assessment

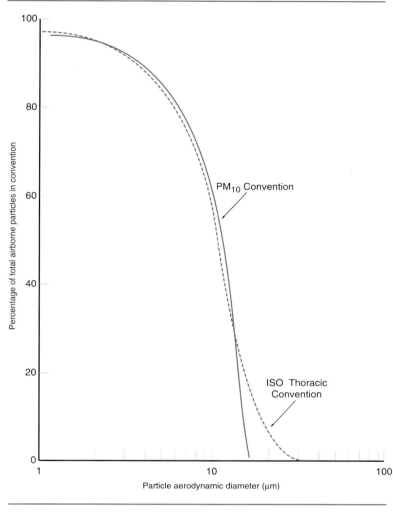

Figure 3.2 PM_{10} and ISO Thoracic Particle Sampling Conventions.

and Management Directive, which has now reached a common position, sets out a number of pollutants to be covered by daughter Directives. These include "suspended particulate matter" and "fine particulate matter (including PM_{10})". Technical work on these has now started within Europe.

There is a US EPA definition of the thoracic aerosol fraction known as PM_{10} (EPA, 1987), which has been systematically measured with validated instruments in the USA for the last 8-9 years. It differs slightly from the ISO definition in that its upper cut off is lower at 16 µm, as shown in Figure 3.2, although in practice this difference is not significant.

The USA is currently giving consideration to implementation of sampling for a finer aerosol fraction, with a median aerodynamic diameter of about 2.5 µm, which may eventually prove to be of greater health significance than PM_{10}. This would separate the finer particles of mostly anthropogenic origin from the coarser particles which include natural components. $PM_{2.5}$ sampling would equate to the high risk respirable fraction defined in IS 7708. Such samplers are already available and are starting to be used in the UK, although not in a routine way.

Ambient particle sampling currently focuses on the mass of the particles rather than their number. Recent suggestions that adverse effects of exposure to ambient particles may be caused by the ultrafine fraction (<0.1 µm) (Seaton et al, 1995), are likely to generate greater research interest in the measurement of particle numbers.

3.2.2 Early Techniques of Health-related Sampling of Particles in the UK

Sampling of ambient airborne particles was introduced into the UK on a routine basis in the first half of this century. In the 1960s the scale of sampling expanded rapidly, in response to the concerns about the health effects of the smogs. A countrywide network of simple samplers operated by local authorities was established to measure the levels of black smoke in the atmosphere. These samplers, shown in Figure 3.3, comprised a downwards-facing funnel into which air was drawn. The sampled particles were collected on a paper filter at some distance (typically 2 m) from the inlet. The quantity of dust was assessed by measurement of light reflectance from the surface of the filter. The black particles prevalent in the smoke from fossil fuels contributed most to the measurements at this time, which was converted to mass units using a calibration curve. This method was adopted by the British Standards Institute as the standard method for measuring airborne particulate matter (BSI, 1969a). It is still used today and is covered by the Sulphur

Figure 3.3 Components of British Black Smoke Sampler Together with its Sampling Efficiency.

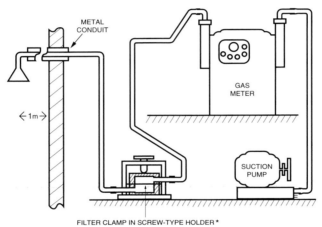

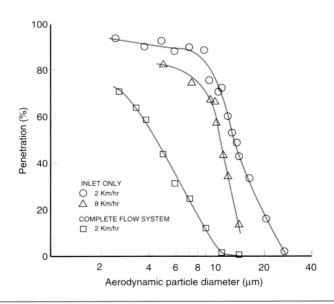

Dioxide and Suspended Particulates Directive (80/779/EEC). It should be noted, though, that the calibration curve in the EC directive differs from that in the British Standard (BS black smoke = 0.85 x EC black smoke). This is a useful reminder that the calibration of the black smoke method depends on the composition of the particles. For current particles, the black smoke method provides an indication of particle concentration but this is both site and season-specific. This can be seen in the comparisons between black smoke and PM_{10} measurements reported in Chapter 6. The sampling performance of the BS system (McFarland and Ortiz, 1984) is also shown in Figure 3.3, where it can be seen to collect particles below 4 µm with a high efficiency. It therefore approximates to the 'respirable' fraction defined in IS 7708.

In the early 1970s, the Government instigated two other national surveys of ambient dust to monitor sulphate aerosol and metals. Two very similar samplers were used in the surveys. The "S-type" for sulphate and the "M-type" for metals. The "M-type" sampler, is shown in Figure 3.4. It comprises a downwards-facing cylindrical hood covering a 37 mm open-face filter holder. The design was influenced primarily by the need to protect the filter from wind and rain. However, this inevitably meant that large particles were not sampled, as can be seen in the results of the wind-tunnel performance tests also shown in Figure 3.4 (Upton and Barrett, 1985). Its performance is strongly windspeed dependent, although at 6 m/s, which is close to the average wind speed in the UK, its sampling efficiency is close to

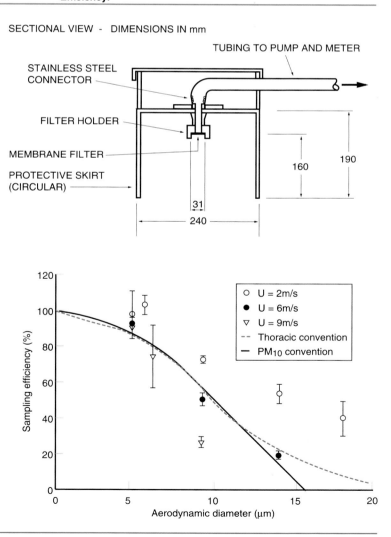

Figure 3.4 Details of M-type Sampling Head Together with its Sampling Efficiency.

that of a PM_{10} sampler. Although the M-type sampler is used in the UK for monitoring compliance with the EC lead directive (82/884/EEC) which sets an annual limit value, its performance characteristics clearly restrict its use as a health-related sampler over short time periods. However it may well be of value as an indicative PM_{10} sampler when the results are averaged over longer time periods, ranging from several months to a year. To support this view, results from monitoring in London over several years with the M-type sampler show annual mean concentrations that are very similar to those being measured with the PM_{10} sampler that forms part of the national network.

3.2.3 Samplers Designed to Monitor Health-related Particle Fractions

The majority of reliable instruments currently available are for the PM_{10} fraction. These generally make use of a validated sampling head to select the PM_{10} fraction of the ambient airborne particles, with the collected particles being analysed in two main ways:

1) Gravimetric, cumulative samplers in which the PM_{10} particles are deposited on a filter over a sampling period of normally 24 hours. The mass of particles collected on the filters is determined by weighing.

Table 3.1 Examples of Gravimetric, Cumulative PM_{10} Particle Samplers for the Ambient Atmosphere Currently Available in the UK.

Name	Flowrate l/min	Filter Diameter mm	Comments
PQ167 Portable PM_{10} Sampling Unit	16.7	47	Uses validated PM_{10} inlet (US EPA Protocol) connected to microprocessor flow-controlled pump. Battery powered - lasts for over 24 hrs using quartz or glass fibre filters
Partisol Model 2000 Air Sampler	16.7	47	Uses validated PM_{10} inlet connected to microprocessor controlled pump. Supplied as stand alone unit or with three additional satellites controlled by hub unit dependent upon wind speed and/or direction conditions
PM_{10} Dichotomous Sampler	16.7	37 (two)	Uses validated SA246b PM_{10} inlet followed by virtual impactor to give two fraction collected on filters; 10 µm - 2.5 µm and < 2.5 µm
PM_{10} Medium Flow Sampler	113	102	Medium flow rate sampler uses teflon or quartz filters, specially used for X-ray fluorescence and other compositional analyses
SA 1200 PM_{10} High Volume Ambient Air Sampler	1130	200 x 250	Standard sampler used in USA for PM_{10} aerosol in ambient atmosphere - high flowrate means short sampling times and large masses for gravimetric and chemical analysis

2) Direct-reading monitors in which the selected PM_{10} particles are either deposited on a filter, with continuous assessment of the change of a property of the filter due to their presence, or passed through an optical sensing region.

The $PM_{2.5}$ fraction can be measured using the same analysers, provided use is made of a $PM_{2.5}$ sampling head.

3.2.3.1 Gravimetric, Cumulative Samplers

These samplers basically comprise an omnidirectional rain-protected entry followed by a size-selective stage (normally an impactor) to separate the PM_{10} particles which are collected on the filter. There are a number of different samplers available, ranging in flowrate from 16.7 to 1130 l/min. A summary of the important features of some samplers available in the UK is given in Table 3.1 and a brief description of the design features and relative performance of the sampling heads is given below.

High flowrate samplers such as the Graseby Andersen PM_{10} Hi-Vol Sampler (see Figure 3.5) have been the mainstay of routine PM_{10} measurements in the USA. The high flowrate of 1130 l/min has the advantage of providing sufficient sample both for gravimetric and chemical analysis over the specified 24 hr sampling period. Hi-Vol samplers (without a complete PM_{10} inlet) were used in the Department of the Environment's national survey of dioxins, PCBs and PAHs in the ambient atmosphere (Coleman et al, 1995). The important features of the PM_{10} sampling head are an omnidirectional narrow slot entry, which samples particles independently of windspeed up to 10 m/s, and a multi-orifice single stage impactor, which allows the PM_{10} fraction to penetrate to a 25 x 20 cm filter. The efficiency of the sampler demonstrates full agreement with the US EPA PM_{10} Convention, as shown in Figure 3.5. The sampler is mains powered with either volume or mass flowrate automatically controlled. A high volume virtual impactor is available to insert between the impactor and the filter to provide the $PM_{2.5}$ fraction of the PM_{10} particles.

A number of low flowrate PM_{10} samplers are available in the UK. They all make use of the SA246b PM_{10} inlet developed and validated by Graseby Andersen. The entry (see Figure 3.6) consists of a flanged cylindrical pipe with a disc rain cap held some distance above. This forms the omnidirectional entry through which particles enter, followed by a single stage impactor which allows the PM_{10} fraction to

Figure 3.5 High volume PM_{10} Sampler Together with its Sampling Efficiency.

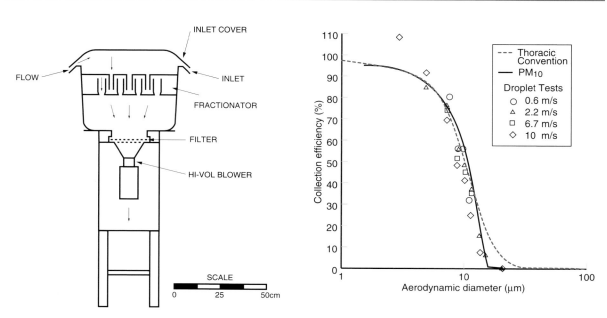

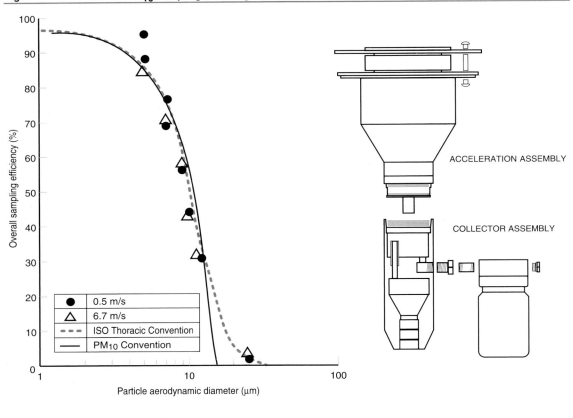

Figure 3.6 Low Volume PM$_{10}$ Sampling Head Together with its Sampling Efficiency.

penetrate to a filter. The sampling efficiency of the entry, which demonstrates full agreement with the US EPA PM$_{10}$ Convention, is also shown in Figure 3.6.

Three devices are included in Table 3.1, each having separate features that makes them to a certain extent complementary. The Casella PQ 167 Portable PM$_{10}$ Sampling Unit uses a specially designed battery-powered pump which lasts for 24 hrs, and a 47 mm diameter quartz or glass fibre filter. This makes the sampler very useful at sites where mains power is not easily available.

The Rupprecht and Patashnick Partisol Model 2000 Air Sampler can either be supplied as a stand-alone unit or with three additional satellite PM$_{10}$ sampling head and filter units. The satellites are connected to the main hub unit (at a maximum distance of about 3 m) through simple air flow lines which are switched via solenoid valves controlled by a user defined sampling programme. This enables four daily samples to be taken without attention. In addition, separate samples can be taken for different wind speeds, wind directions and at specified times, thereby providing useful evidence for source apportionment studies and for correlation with meteorological conditions. For all units the PM$_{10}$ head can be replaced with a small cyclone inlet which has a 50% penetration at 2.5 μm. The sampler runs off mains power only.

Finally, the Graseby Andersen PM$_{10}$ Dichotomous Sampler separates particles into two distinct fractions: 2.5 to 10 μm and < 2.5 μm. This is achieved by following the PM$_{10}$ inlet with a virtual impactor with a 50% penetration at 2.5 μm, and collecting both fractions on 37 mm diameter teflon membrane filters. This gives some size distribution information of particles within the PM$_{10}$ fraction, and provides separate samples for chemical species studies.

3.2.3.2 Direct-reading Monitors

For these instruments, sampling and analysis is carried out within the instrument, and the concentration can be obtained almost immediately. Like the low flowrate cumulative samplers described above, these instruments generally use the validated SA 246b PM_{10} inlet to select the PM_{10} particles which either deposit on a special filter stage or penetrate into a particle sensing region. Instead of direct weighing the presence of the particles either on the filter or in the sensing region gives rise to a property change, which can be related by calibration to the mass of particles present. A number of different instruments are available and these may be classified into three main categories: optical, resonance oscillation, and beta particle attenuation. A summary of the main features of the direct-reading instruments for health-related purposes is given in Table 3.2, and the principles of operation are described below.

Optical

These instruments employ the interaction between airborne particles and visible light in a sensing region, and generally their response is dependent upon the size distribution, shape and refractive index of the particles. They therefore require calibration to give results in terms of mass or number concentrations. This calibration only holds provided that the nature of the particles does not change and hence measurements obtained in ambient air are open to considerable uncertainty. Two instruments are currently available.

The DataRAM Portable Real-Time Aerosol Monitor uses a near infrared source and forward angle scattering to detect the concentration of particles in the range 0.1 to 1000 µg/m³. It uses a scaled-down version of the SA 246b PM_{10} inlet to select the PM_{10} fraction at a flowrate of 2 l/min, although this has not been experimentally validated. The impactor stage can be replaced to give a nominal 50% penetration at 2.5 µm. The instrument is powered by a 6 V lead acid battery which lasts for 20 hours, making it useful at sites where mains power is not easily available. It is calibrated with AC fine test dust (mass median aerodynamic diameter 2 to 3 µm, standard deviation 2.5), and may need on-site calibration to ensure valid results in terms of mass concentration.

The Grimm Stationary Environmental Dust Analyser 1.200 comprises two samplers in one unit. The "reference" unit uses a standard 16.7 l/min PM_{10} inlet connected to a 25 mm diameter glass fibre filter. This is essentially a standard gravimetric sampler as previously described, with the concentration determined by weighing the filter. However an airflow splitter is used to extract air at 1.26 l/min downstream of the PM_{10} head, this sub-sample being fed into a light scattering optical particle counter. The optical counter uses peak height analysis to separate the particles into 8 channels in the range 0.3 to 15 µm. The mass concentration within these fractions is determined using calibration factors which can be determined from an in-built back-up filter (although a long sample period is required to collect sufficient sample on the back up filter with the low flow rate used). The optical particle counter is also available as a stand-alone unit as a nominal PM_{10} sampler. In this form it is particularly attractive because of its small size. It is a potentially versatile tool that provides both number and (nominal) mass size distributions. However, its performance as a mass sampler depends crucially on the calibration, which is based on integration over a long period and over all size fractions. It is not clear, yet, how variable the calibration factor is with time (hour by hour) and from one size fraction to another. Until these issues are addressed, the results from the optical unit must be treated with caution.

Oscillating Microbalance

The frequency of mechanical oscillation of an element such as a tapered glass tube is directly proportional to the mass of the tube. Change in effective mass of the tube, such as that due to deposition of particles on the surface of a filter at the free end of the tube, is reflected in a change in its resonant frequency.

This is the principle of operation behind the Rupprecht and Patashnick Tapered Element Oscillating Microbalance (TEOM), as shown in Figure 3.7. A standard 16.7 l/min PM_{10} inlet selects the PM_{10} particles which pass through a flow splitter in which 3

Table 3.2 Examples of Direct-Reading Monitors for PM_{10} Particles in the Ambient Atmosphere.

Name	Measurement Technique	Flowrate l/min	Particle Fraction	Concentration Range µg/m³	Precision µg/m³ 1 hour	Precision µg/m³ 24 hours	Comments
TEOM Series 1400a Ambient Particulate Monitor	Tapered element oscillating microbalance	16.7 thro inlet with 3 thro' filter/detector	PM_{10} with possible 2.5 or 1.0 µm	0.06 - 1500	1.5	0.5	Only direct-reading monitor in which output directly related to mass. Employed as particle monitor at national Automatic Urban Network Sites.
W & A Beta Gauge Automated Particle Sampler	Attenuation of beta rays by particles collected on a filter	18.9	PM_{10}	$4 - 10^4$	4	0.1	One of a number of filter tape based beta gauges - measurement cycle 1 hour.
Airborne Particle Monitor APM1	Attenuation of beta rays by particles collected on a filter	15 - 30	PM_{10} (non validated)	$2 - 10^7$	56	2	Cassette system with 30 filters in sequential loader. Integrity of each sample maintained for compositional analysis.
GRIMM Model 1.104 Dust monitor	Light-scattering photometer	16.7 thro' inlet 1.26 thro' detector	PM_{10} with EPA validated head	$1 - 5 \times 10^4$ indicated	Not given	Not given	Optical particle counter with in-built filter for on-site calibration, as response may be dependent upon refractive index and size of particles.
DataRAM Portable Real-Time Aerosol Monitor	Light-scattering photometer	2	PM_{10} or $PM_{2.5}$ (non validated)	$0.1 - 10^3$ indicated	1.0	Not given	Optical device calibrated with AC fine test dust. May need on-site calibration to give reliable mass measurements as response dependent upon refractive index and size of particles. Entry dependent upon windspeed.

Figure 3.7 Schematic Diagram of Rupprecht and Patashnick TEOM Ambient Aerosol Monitor.

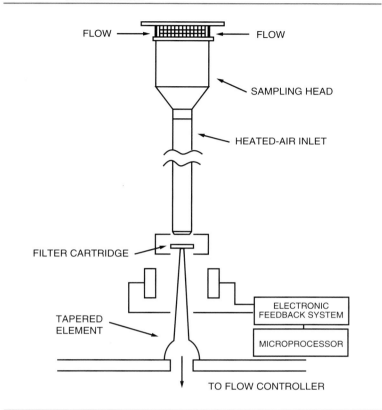

l/min passes through a 16 mm diameter filter connected to the top of the narrow end of a hollow tapered glass tube. As the particles collect on the filter, the tube's natural frequency of oscillation decreases. The change in this frequency is directly proportional to the added mass. The inlet including the sensing system is kept at a steady 50°C to drive off any sampled water droplets. The instrument is microprocessor controlled and the mass concentration values are updated every 13 seconds with average concentrations provided every 30 minute or every hour. This instrument is the particulate monitor chosen for the Department of the Environment's Automatic Urban Network sites and has been used by many organisations both in the UK and worldwide. Some concern has been expressed about the potential loss of volatile material at the stable temperature of 50°C (Weiner, 1995).

Beta Particle Attenuation

This involves the measurement of the reduction in intensity of beta particles passing through a dust-laden filter or collection substrate. The change in attenuation reflects the rate at which particles are collected on the filter and hence the concentration in the sampled air. The mass of particles collected is calculated from a knowledge of their mass absorption coefficient. The instrument is calibrated using the mass absorption coefficient for quartz. Measurements with ambient aerosols suggest the mass absorption coefficient may in practice vary by up to ±20%. The beta attenuation method is therefore inherently less accurate than gravimetric methods. Two main types of instrument have been developed: one using filter tape to collect the particles, and the other using a stack of conventional filters in a sequential loader.

There are many tape-based beta particle attenuation devices available, and details of just one typical example are given here.

The Wedding & Associates Beta Gauge Automated Particle Sampler uses their version of the low volume, validated PM_{10} inlet at 18.9 l/min to select the PM_{10} particles which are deposited on to either a glass fibre or PTFE filter tape. A beta particle attenuation system comprising a 3.7 MBq ^{14}C source and a fast response silicon semiconductor detector is used to detect the presence of dust on the filter with a one hour measurement cycle.

The Elecos Airborne Particle Monitor APM-1 is an interesting alternative to the tape based beta particle attenuation monitors. Instead of the filter tape, it deposits PM_{10} particles onto one of thirty 47 mm diameter filters held in a sequential loader system. This prevents violation of the integrity of the deposited particles enabling subsequent unambiguous gravimetric and chemical analyses to be made.

Beta particle monitors have been widely employed in a number of European countries. Further evaluation of Beta particle monitors by setting them up alongside a TEOM is however desirable.

3.2.3.3 Health-related Samplers for Coarse Particles

Particles as large as 100 μm can enter the human nose and mouth during breathing. Whilst the occurrence of these large particles may be rare, they can occur close to industrial processes and during periods of high windspeeds. Once inhaled they will deposit in the nasopharyngeal region and if toxic (such as lead, radioactive particles, etc) may enter the blood system there or in the gut.

The relevant ISO health-related fraction is the inhalable fraction, for which there is currently no commercially available instrument. The old "total suspended particulate" (TSP) was effectively defined by the "high-volume" sampler used to measure this parameter. The performance of this well-known device is both orientation and windspeed dependent, with its efficiency falling well below the inhalable aerosol convention, especially with large particles (Wedding et al, 1977).

Recent work has been carried out to develop a sampler to match the requirements of the ISO Inhalable aerosol fraction (Mark et al, 1990). It uses a single orifice that rotates through 360° (see Figure 3.8) to make it omnidirectional. A protective canopy protects it from rain and snow. The efficiency, as tested in a large wind tunnel, at different wind speeds is shown in Figure 3.8 (Mark et al, 1990, 1994).

3.2.4 General Sampling Procedures to Monitor Health-related Particle Fractions

3.2.4.1 Selection of Samplers for PM_{10} Monitoring

The choice of samplers will be influenced by the timescale required for the measurement. The standard proposed by the Expert Panel on Air Quality Standards is 50 μg/m³ as a rolling 24-hour average. This requires measurements to be made to a 1-hour timescale, with the 24-hour average being updated every hour. However, measurements of 24-hour concentrations over a fixed time frame will still provide a useful indication of air quality in relation to the proposed standard, but will underestimate the frequency and magnitude of exceedences. There is a range of samplers that meet the 1-hour or fixed 24-hour time frames, as indicated in Tables 3.1 and 3.2. The majority are relatively expensive, in the range £5,000-£20,000. Unfortunately there is no very cheap sampler available, as is the case with nitrogen dioxide, where the diffusion tube is now a widely used option to provide indicative results. The nearest equivalent for sampling of particulate matter is the 'M Type' sampler, which costs around £1,500. Its sampling characteristics make it a reasonable indicative sampler for long-term (eg annual average) PM_{10} concentrations, although other more expensive samplers with true PM_{10} inlets are preferable.

The TEOM is the sampler used in the UK Automatic Urban Network. It is relatively easy to operate once set up and provides concentrations resolved to 1-hour. It is however the most expensive sampler. Beta particle gauges are somewhat cheaper and the latest models are claimed to be capable of providing data resolved to 1-hour. As discussed above, their accuracy is probably not as good as the gravimetric samplers. The only other instruments capable of

Figure 4.2 Size Distributions of Particulate Emissions from Diesel Engined Cars (USEPA, 1985).

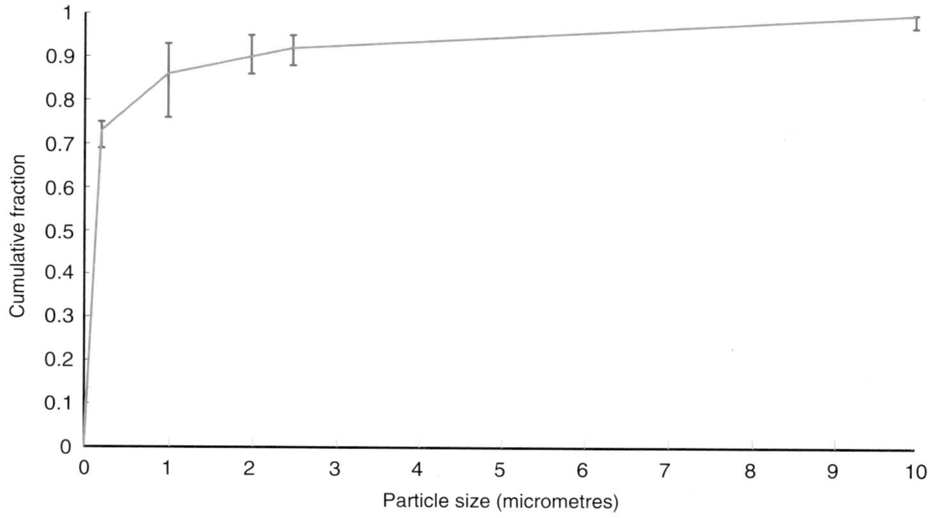

Note: Error ranges are 95% confidence intervals.

accumulation mode of about 0.05 to 1 µm diameter. (See Glossary for definition of terms).

The official European test procedure collects material on a filter through a heated line and so may not measure exactly the mass of particulate matter actually emitted into the atmosphere. Research shows that the sizes of diesel particles are virtually all less than 10 µm (Table 4.1, Figures 4.1 and 4.2) with particulate matter from the use of leaded petrol slightly larger (USEPA, 1990). The factors in this table were applied to the emission rates quoted in this section to obtain the emission rates of particular size fractions.

PM_{10} emission rates for existing vehicles are those detailed in *UK Petrol and Diesel Demand* (Gover et al, 1994) which result from a detailed survey of the literature. The most detailed information is available for diesel cars and Table 4.2 shows the average emission rates of particulate matter on different types of road.

Less is known about emissions from other diesel vehicles, such as buses and coaches. Far fewer measurements have been made on large vehicles and some of these factors, especially for buses, have been estimated on the basis of measurements made on vehicles with similar engines and a few direct measurements. Table 4.3 shows the factors for other diesel vehicles.

There is little available data for petrol engine emissions of particulate matter. The factors in Table 4.4 are the result of a literature survey (Gover et al, 1994; USEPA, 1985) which included some US data which showed the highest mass emission rates for old petrol engined vehicles with leaded gasoline. The range indicates that found in the literature.

Emission limits for particulate matter mass from diesel vehicles will fall in the future due to a series of European Union Directives (91/542/EEC, 94/12/EEC, 93/59/EEC). Table 4.5 indicates the reductions in the

Table 4.2 Particulate Matter Emission Factors from IDI Diesel Car Without Catalyst (Regulation ECE 15.04 (83/351/EEC)) - g/km.

Road Type	Engine Temperature	Engine Size		
		Small (<1.4 l)	Medium (1.4-2 l)	Large (> 2 l)
Urban	Cold	0.24	0.34	0.36
Urban	Hot	0.12	0.17	0.18
Minor Rural	Hot	0.10	0.13	0.15
Major Rural	Hot	0.09	0.11	0.13
Motorway	Hot	0.15	0.20	0.22

Table 4.3 Particulate Matter Emission Factors for Diesel Road Vehicles - g/km (as PM_{10}).

		Road Type			
		Urban	Minor Rural	Major Rural	Motorway
LGV	IDI non-catalyst	0.28	0.23	0.21	0.29
Small HGV	3.5 - 7.5 t GVW	1.90	1.17		0.83
Medium HGV	7.5 - 17 t GVW	1.53		0.95	1.08
Large HGV	17 - 38 t GVW	1.52		0.91	0.71
Mini-bus	up to 16 seats	0.40		0.36	
Midi-bus	17-35 seats	1.90		1.17	
Bus	over 35 seats	1.60		1.40	
Coach	over 35 seats	1.52		0.91	0.71

Table 4.4 Emission Factors for Total Particulate Matter from Petrol Engined Vehicles - g/km (as PM_{10}).

	Emissions	Range
Car non-catalyst leaded	0.06	0.02 - 0.16
Car non-catalyst unleaded	0.02	0.02 - 0.03
Car three-way catalyst	0.01	0.01 - 0.02
LGV non-catalyst leaded	0.08	0.05 - 0.11
LGV non-catalyst unleaded	0.04	0.03 - 0.05
LGV three-way catalyst	0.02	na

emission rates implied by these Directives. These are based on the Directive test values assuming that on-road emissions are at the maximum levels specified by the Directives, and hence the actual emission rates may be different in practice.

The introduction of unleaded petrol and three-way catalysts has resulted in a decrease in the emission rates of particles from petrol engines but this reduction was not the result of a specific particulate emission limit regulation. Emissions reduced with the introduction of unleaded fuel, and a further reduction with the use of three-way catalysts was the result of emission limits for NO_x, VOC and CO.

Data on particulate matter emissions from large diesel engines, trucks, buses and coaches are sparse. In particular the contribution of these vehicle types in the past is poorly quantified. Often, measurements were made on engine test beds rather than whole vehicles on the road or on a chassis dynamometer. Also, while vehicle emissions may increase as the vehicle ages

Table 4.5 Past and Future Reduction in Regulated Particulate Matter Emission Rates.

Date of introduction	Petrol with three-way catalyst	Diesel Car IDI	Diesel LGV	HGV <85kW	HGV >85kW
Pre 1993	100	100	100	100	100
31/12/92		56			
1/10/93				46	46
1/10/94			56		
1/1/96		32			
1/10/96				12	19
1/10/97			44*		

Normalised emission rates. Pre-1993 = 100
* This is a weighted average as different rates apply to different weight classes

Table 4.6 Exhaust Emissions of PM_{10} from Road Traffic (ktonnes).

		1970	1975	1980	1985	1990	1995	2000	2005	2010	2015	2020	2025
Cars	Petrol	9.0	10.7	12.6	14.7	14.8	10.3	4.5	4.1	4.2	4.4	4.7	5.0
	Diesel	0.5	0.6	0.7	1	2.1	4.5	5.1	4.8	5.1	5.7	6.1	6.5
	All Cars	9.5	11.3	13.4	15.7	16.9	14.8	9.6	8.9	9.3	10.1	10.8	11.5
LGV	Petrol	1.2	1.4	1.5	1.6	1.4	0.2	0.1	0.1	0.1	0.1	0.2	0.2
	Diesel	1.1	1.2	1.3	1.4	4.3	8.3	6.2	5.8	6.6	7.5	8.5	9.6
	All LGV	2.3	2.6	2.8	3.0	5.7	8.6	6.3	5.9	6.7	7.6	8.6	9.8
HGV	Large	3.3	3.8	5.0	5.7	8.2	7.1	3.3	2.2	2.4	2.6	2.9	3.2
	Small	21.5	22.8	23.2	22.1	25.5	20.1	9.3	6.2	6.7	7.4	8.2	9.1
	All HGV	24.8	26.6	28.2	27.8	33.7	27.2	12.6	8.4	9.1	10.0	11.1	12.2
Buses and Coaches		5.3	4.8	5.2	5.3	6.7	6.3	4.2	3.1	2.0	1.4	1.3	1.3
Motorcycles		0.4	0.4	0.7	0.6	0.6	0.4	0.4	0.4	0.4	0.5	0.5	0.5
All Diesel		31.6	33.2	35.4	35.5	46.8	46.3	28.1	22.1	22.8	24.6	26.9	29.7
All Petrol		10.6	12.5	14.8	16.9	16.7	10.9	5.0	4.6	4.7	5.0	5.4	5.7
Total		**42.1**	**45.7**	**50.2**	**52.4**	**63.6**	**57.2**	**33.1**	**26.8**	**27.5**	**29.6**	**32.3**	**35.4**

Note: Assumes diesels comprise 20% of new car sales.

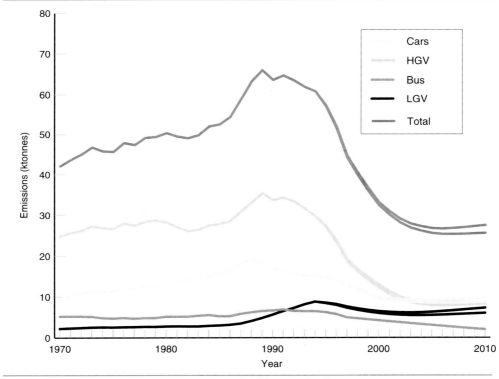

Figure 4.3 UK Road Transport Emissions of PM_{10} (Diesel Car Sales 20%).

Note: The split lines in future projections correspond to low and high forecasts of traffic.

this has not been well quantified. In the figures quoted in this Chapter any effect of vehicle wear and tear has not been included.

Table 4.6 and Figure 4.3 show the emission estimates that result from these emission factors. The future vehicle usage assumed in the calculation is the mean of the upper and lower projections from the Department of Transport's National Road Traffic Forecasts (these are predictions of road traffic up to

2025). The introduction of low sulphur diesel in 1996 will reduce particulate matter emissions from existing diesel engines by up to 13% for HGV and have a smaller, but beneficial effect upon emissions from light duty vehicles. Diesel cars are assumed to comprise 20% of new car sales, close to the percentage achieved in 1994/95.

4.2.2 Road Transport - Other Emissions

Particulate matter is also emitted from tyre wear and brake wear. Information from the USA gives an indication of the importance of these sources.

4.2.2.1 Tyre Wear

The USEPA gives a factor of 0.002 g/mile for particles <10μm (0.0012 g/km) for tyre wear from cars. As the particles are produced by attrition processes, particle sizes are probably larger than for exhaust emissions, and so emissions of particles less than 2.5μm are likely to be negligible. Assuming that HGVs have larger emission rates (scaling as the number of wheels) and that these figures apply to the UK this gives the emission rates shown in Table 4.7.

4.2.2.2 Brake Linings

The USEPA gives a factor of 0.0128 g/mile (0.00795 g/km) of particles from wear of brake linings on cars. Table 4.1 shows that 98% of these emissions are less than 10μm. Again, assuming that emissions from large vehicles are larger than from cars (emissions are scaled according to the number of wheels) and that these US figures apply to the UK, emission estimates appear in table 4.8.

Table 4.7 Annual Tyre Wear Emissions of PM_{10} (ktonnes) in the UK.

	1970	1980	1990	1993
Tyre Wear Emissions (ktonne)	0.31	0.39	0.57	0.57

Table 4.8 Annual Brake Wear Emissions of PM_{10} (ktonnes) in the UK.

	1970	1980	1990	1993
Brake Wear Emissions (ktonne)	1.95	2.61	3.78	3.80

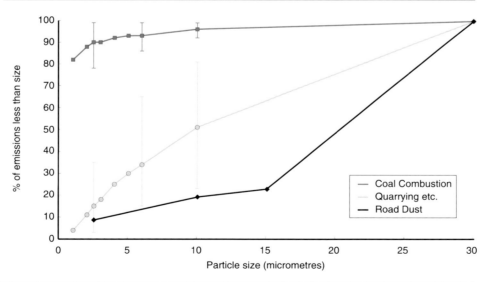

Figure 4.4 Typical Cumulative Particle Size Distributions for Particles from Coal Combustion, Quarrying and Road Dust (USEPA, 1990).

Note: Error ranges are 95% confidence intervals.

Table 4.13 - PM$_{10}$ Sources in Order of Decreasing Reliability of Emission Estimate.

Source	Comments
Diesel Cars	Based on a number of detailed measurements
Diesel LGV and HGV	Based on a few measurements
Diesel Buses and Coaches	Only a few measurements. For some types the factors are extrapolated from similar vehicle types.
Petrol Cars	Only a few measurements. The factors chosen are uncertain especially for cars without catalysts
Other Mobile Sources	Little data and uncertain activity data.
Stationary Combustion	Only a few measurements. Some of the PM$_{10}$ factors have to be interpreted from smoke or TSP data
Industrial Processes	Few data of questionable applicability. Order of magnitude estimates possible.
Mining and Quarrying	Few data of questionable applicability. Order of magnitude estimates possible.
Construction	Only USA data which is not strictly appropriate for the UK. Order of magnitude estimates possible.

- In the absence of any other information it is assumed that emissions from non-fuel combustion stationary sources will continue into the future at their current rates. Changes in PM$_{10}$ into the future will depend primarily on changes in two sectors; fuel combustion and road transport.

- Road transport emissions of PM$_{10}$ will change as result of two factors: increasing road usage and the implementation of stricter emission limits on new vehicles. The National Road Transport Forecasts (NRTF, DoT, 1994) have been used as the basis of predictions of the growth of vehicle kilometres and vehicle fleets.

- Future European emissions regulations that have already been agreed to are included. Limits for HGVs come into force in 1995/6 (91/542/EEC). New car limits will come into force in 1996 (94/12/EEC) while limits on diesel vans will come into force in 1997 (93/59/EEC). New, Stage III limits for emissions from vehicles are planned to be introduced by the EU in 2000. These are, however, still the subject of discussion and cannot be included in the projections. It is assumed that diesel cars will account for 20% of all car sales each year from 1995.

The make-up of the vehicle parc, and in particular the percentage of diesel vehicles in the car fleet is an important determinant of the emissions from road transport. In our Second Report (QUARG, 1993), this Review Group showed that future reductions in particulate matter emissions effected by tighter exhaust emission regulations could be largely lost if the percentage of diesels in the car fleet approached 50% as in some other European countries. Hence the predictions are highly sensitive to the market share of diesel cars.

Figure 4.3 shows the latest projection of road transport emissions of PM$_{10}$ in the UK, whilst Figure 4.8 shows the road transport emissions in urban areas only. As this follows the NRTF, it may overestimate emissions for London, where traffic growth is now well below the national average. Emission estimates for stationary fuel use are based on estimates of fuel use published by the DTI (DTI, 1995). The scenario based on 'central growth, low fuel prices' (DTI, 1995) is used. While the amounts of fuel used change, specific emission rates are assumed to be constant. Figure 4.9 shows national emission estimates until 2010.

Estimates of future emissions in urban areas depend on the traffic growth estimates for those urban areas. Such estimates are not currently published, and each urban area is likely to be different. However, it is clear that road transport will continue to be dominant in urban areas until after 2010.

4.6 UNCERTAINTIES

Emission inventories are estimates of emissions of specific pollutants from a given area over a specified time period. While measured data is used where possible, it is also necessary to include estimates of emissions from some categories of source because it is impossible to measure all the sources all the time. Additionally it cannot be stated with certainty that all the sources that exist have been considered. In practice, the major sources are included as they are well known and any 'missing' emissions are small.

The estimates are generally based on a few measurements with the assumption that these results are typical of the behaviour of all similar sources.

The emission estimates of PM_{10} presented in this Report are based where possible on measurements of PM_{10} emissions, although in some instances measured total particulate matter emissions have been converted to PM_{10}. While there have been a number of investigations of particulate mass emission from diesel-engined motor vehicles, this is not true of many fugitive sources. Stationary fuel combustion falls between these extremes. The figures presented above are indicative estimates only for some sources, particularly construction, quarrying and industrial processes. Given the lack of knowledge about these emissions it is not possible to give quantitative indications of the accuracy of the emission estimates. It is however possible to give an indication of the reliability of the estimates, and these are presented in Table 4.13.

A joint research programme sponsored by the Department of the Environment, Department of Transport and the motor industry is currently investigating size-fractionated mass and number emissions of particles from petrol and diesel vehicles, and should enable more accurate estimates of emissions from road vehicles to be made in the future.

4.7 BIOLOGICAL PARTICLES

The biological content of airborne particulate matter has been an important area of scientific interest since the nineteenth century. Transmission of hospital infections and pathogenic crop diseases are known to be related to the varied airborne dispersal mechanisms of micro-organisms, and considerable research has been undertaken within these fields.

The biological particles present in ambient air, including pollen grains, fungal spores, bacteria and viruses, have all been implicated in a wide variety of health issues. Clearly, allergenic response to certain pollen species presents an annual misery for millions of hayfever sufferers across Europe. Also, occupational diseases such as Farmer's Lung and Toxic Organic Dust Syndrome are known to be related to extended exposure to high concentrations of fungal spores found in certain working environments. Direct infection by such organisms as Legionella bacteria from air conditioned buildings has also held a high media profile, although with respect to a relatively small number of cases. High episodic levels of spores released by heavy summer rain have been related to ten fold increases in asthma admissions (Packe and Ayres, 1986). Although the significance of natural allergens for asthmatics has been well documented, little is currently known of the relative mass composition of the species-specific or general biological aerosol with respect to the detailed determination of urban PM_{10} concentrations.

Table 4.14 Atmospheric Micro-organisms and their Relative Sizes

Micro-organism Type	Mean size (µm)
Pollens	10-100
Fungal Spores	5-40
Bacterial Spores	0.3-20
Viruses	0.003-0.5

Biological components associated with airborne particulates have historically been assessed by traditional microbiological growth techniques (agar plates) or by optical (microscope) analysis. The typical sizes of different types of micro-organism appear in Table 4.14. Many sampling techniques are now understood to be far from quantitative, and in addition, many detection methods are significantly biased towards some organisms and against others. This lack of qualitative and quantitative information suggests data from past research is often not representative of the real environment, nor easily comparable to other research. Data on the sources and variability of bioaerosols in the atmosphere will therefore be dependent upon these sampling and analytical issues. In addition, the biological viability of any given species in the atmosphere will depend upon its tolerance of UV radiation, relative humidity and temperature fluctuations, although the importance of the physical aerosol may not necessarily be related to its biological activity.

The terrestrial rather than marine environment is generally considered to be the major source of airborne micro-organisms (Roffey et al, 1977). Vegetation and general farming activities, including animal husbandry and arable cropping are major contributors to the ambient bioaerosol, although release mechanisms will depend strongly on meteorological conditions. Urban sources of bacteria have been estimated to be up to an order of magnitude greater than corresponding rural regions (Bovallius et al, 1978). However seasonal and regional activities, such as wheat harvesting, may reverse the source apportionment of specific species. More easily defined anthropogenic sources include sewage treatment works, landfill sites and power generation cooling towers.

Although variability in composition will depend heavily on seasonal and diurnal cycles, specific species viability and release may be greatly influenced by the relative impact of rainfall, high wind speeds, and fluctuating humidity. Overall it has been suggested that bacteria are associated mostly with springtime, yeasts during the summer, and fungal spores in the winter. Pollens are generally greatest during warmer months, with grasses normally peaking in late June and early July. Tree pollens tend to peak earlier in the spring with Platanus pollens highest in April and Betula during May. Diurnal pollen studies have found maximum release generally occurs during late afternoon (Norris-Hall and Emberlin, 1991). Clearly, variability between differing years may be great, and in addition, as little as 1 mm rainfall will rapidly remove the majority of pollens from the atmosphere, leading to large hourly and daily fluctuations. Qualitatively, bioaerosol composition analysed between 1958 and 1961 in Cardiff found that mean pollen levels represented just 2% by number of the bioaerosol content of the atmosphere (Knox, 1979), the remaining components being made up of bacteria and fungal spores. Other quantitative studies have determined fungal spores peaking at over 50,000/m^3 in July, with lowest recordings around 10,000/m^3 in May. Single bracken fronds have been found to release up to 300 million spores during the late autumn. One determination of pollen release from cultivated grass estimated 210 kg per hectare during the year (Knox, 1979).

With the increased awareness of the importance of airborne particulates, it is perhaps surprising that relatively little is known of the importance or relative mass composition of the biological component of ambient air in relation to the PM_{10} levels routinely measured in urban areas. Ongoing development of molecular DNA techniques may soon be able to provide more reliable qualitative and quantitative data on the biological composition of ambient aerosol. Further assessment of the importance of these parameters can then be considered in relation to other current research into urban air quality and health.

4.8 KEY POINTS

- *Road transport is a significant source of primary particulate matter emissions, especially in urban areas. Nationally it accounts for about a third of emissions of PM_{10}; in London this rises to over 80%.*

- *Stationary combustion sources are significant contributors to particulate matter emissions on a national scale. They account for about one half of the PM_{10} emissions. However in cities this contribution may become very small; in London it accounts for about 5% of all the emissions, but is larger in Belfast.*

- *There are a number of industrial processes that can emit large quantities of PM_{10}. While this may dominate emissions locally they do not contribute a large fraction of the emissions nationally. Their impact on urban inventories is generally also small due to their being sited out of town.*

- *Primary national emissions of PM_{10} are predicted to fall by about one quarter between 1995 and 2010.*

- *Future emissions of PM_{10} in cities will depend crucially on the amount of diesel fuel consumed and the proportion of diesel vehicles in the fleet.*

- *Emissions estimates for PM_{10} from different source categories vary greatly in their reliability. The most reliable are from diesel cars and are based on a number of detailed measurements. Contributions from some other sources such as mining and quarrying, and construction are subject to great uncertainty.*

- *Biological particles comprising pollens, fungal spores, bacteria and viruses are widespread. Whilst their number concentrations have been evaluated, their contribution to particle mass is not known.*

REFERENCES

Ball DJ and Caswell R (1985) **Smoke from Diesel Engined Road Vehicles**, Atmos Environ, 17, 169-183.

Bovallius A, Bucht B, Roffey R and Anas P (1978b) **Three Year Investigation of the Natural Airborne Bacterial Flora at Four Localities in Sweden**, Appl Environ Microbiol, 35, 847-852.

Department of the Environment. (1992) **The UK Environment**, HMSO, London 1992.

Department of the Environment (1994) **Sustainable Development. The UK Strategy**, HMSO, London 1994.

Department of the Environment (1995) **Managing Demolition and Construction Wastes. Report by Howard Humphreys & Partners**, HMSO, London 1995.

Department of Trade and Industry (1995) **Energy Projection for the UK. Energy Paper 65**, HMSO, London 1995.

Department of Transport (1995) **Transport Statistics GB, 1994**, Department of Transport, London.

Eggleston HS et al (1993) **CORINAIR Working Group on Emission Factors for Calculating 1990 Emissions from Road Traffic Volume 1: Methodology and Emission Factors**, EEC, Brussels & Luxembourg 1993.

Gillham CA, Couling S, Leech PK, Eggleston HS and Irwin JG (1994) **UK Emissions of Air Pollutants, 1970-1991**, WSL Report LR 961, Stevenage, Herts.

Gover MP et al (1994) **UK Petrol and Diesel Demand - Energy and Emissions Effects of a Switch to Diesel**, ETSU Report, Harwell 1994.

Hjellbrekke AG, Lovblad G, Sjoberg K, Schaug J, Skjelmoen JE (1995) EMEP Data Report 1993, Part 1: Annual Summaries, EMEP, Norway.

HMIP **Chemical Release Inventory** 1994.

Knox R.B (1979) **Studies in Biology, No. 107; Pollen and Allergy**, Edward Arnold Publishers Ltd.

LRC (1993) **Energy Use in London**, London Research Centre, London.

National Power (1994) **Environmental Performance Review**, National Power.

Norris-Hall J and Emberlin J (1991) **Diurnal Variation of Pollen Concentration in the Air of North-Central London**, Grana, 30, 229-234.

Packe GE and Ayres JG (1986) **Aeroallergenic Skin Sensitivity in Patients with Severe Asthma during a Thunderstorm**, The Lancet, i, 850-851.

PowerGen (1994) **Environmental Performance Report**, PowerGen.

PowerGen (1995) **Environmental Performance Report**, PowerGen.

Precision Research (1995) **Off-road Vehicle Emissions Equipment Ownership and Usage**.

QUARG (1993) **Diesel Vehicle Emissions and Urban Air Quality**, QUARG, London.

Roffey R, Bovallius A, Anas P, and Konberg E (1977) **Semicontinuous Registration of Airborne Bacteria at an Inland and Coastal Station in Sweden**, Grana, **16**, 171-177.

USEPA (1990) **Air Emissions Species Manual, Volume II: Particulate Matter Species Profiles, 2nd Edition**, EPA-450/2-90-001b, USEPA Research Triangle Park, NC, USA.

USEPA (1985) **Compilation of Air Pollutant Emssion Factors 4th Edition, Volume II: Mobile Sources**, USEPA, Ann Arbor, USA.

USEPA (1995a) **Compilation of Air Pollutant Emission Factors 5th Edition, Volume I: Stationary Sources**, USEPA, RTP North Carolina, USA.

USEPA (1995b) **Factor Information Retrieval System**, USEPA, RTP North Carolina, USA.

5 Sources and Concentrations of Secondary Particulate Matter

5.1 INTRODUCTION TO SECONDARY PARTICULATE MATERIAL

In this section, attention is directed to that fraction of the suspended particulate material in the particle size range of less than 10 micrometres (µm) in diameter, which was not directly emitted into the atmosphere in particle form. A significant fraction of the PM_{10} consists of particles containing material which was originally present in the gas phase in the atmosphere, but has been subsequently taken up into the particulate phase. The term secondary particulate material refers to this material, originally gaseous but present as an intimate component of the PM_{10}.

Reviews of the many thousands of measurements of the particle size distributions of the suspended particulate matter in urban atmospheres have been synthesised to yield idealised relationships between the number of particles, their superficial area, volume and mass, and their particle diameters. Figure 5.1 shows the idealised particle size distribution of urban aerosols in the USA, in the form of three curves showing the number-, area- and volume-size distributions, taken from Whitby (1978).

Figure 5.1a shows the fraction of the number of particles in each size range in the idealised urban aerosol. The curve shows a predominant single peak with a shoulder to higher particle sizes. This number distribution peaks at about 0.013 µm (13 nm) with a total number density of about 100,000 particles/cm³. Particles in the size range less than 0.1 µm are said to be in the nucleation mode and include the Aitken nuclei.

Nucleation mode particles have been emitted into the atmosphere as primary particles by combustion sources (primary particulate material), both stationary and mobile. These particles include elemental carbon from diesel vehicles and flyash from pulverised coal-fired power stations. However, some nucleation mode particles are formed by condensation of gaseous material through gas-particle conversion processes (secondary particulate material). Few vapours can readily form totally new nucleation mode particles directly themselves but sulphuric acid vapour, formed by the photochemical oxidation of sulphur dioxide, exhibits this ability. In urban areas, nucleation mode particles are the most numerous of all the particles.

Figure 5.1b shows the fraction of total surface area of the particles in each size range in the same idealised urban aerosol as Figure 5.1a. In this case the curve shows a strong single peak with shoulders to both higher and lower particles sizes. The area distribution peaks at about 0.1 µm with a total area of about 500 µm² /cm³. Particles in the size range 0.1 to 1 µm are said to be in the accumulation mode.

Few accumulation mode particles were actually emitted into the atmosphere in this particle size range. Because this mode contains the bulk of the surface area of the particles, see Figure 5.1b, these particles offer the largest target area for adsorbing other aerosol particles (coagulation) or for adsorbing gaseous material. The process of coagulation reduces the number of particles. The process of adsorption of gaseous material from the atmosphere onto pre-existing nucleation or accumulation mode particles leaves the number of particles unchanged but causes the area, volume and mass of the suspended particulate material to grow rapidly. The accumulation mode is therefore the aging region for all small particles because growth is rapid and loss processes are at a minimum at 0.1-0.3 µm. Thus, particles originally present as nucleation mode particles tend to accumulate in the accumulation mode. Typically there might be about 10,000 particles per cm³ in the accumulation mode in an urban area. Coarse and nucleation mode particles make a small contribution to the total superficial area of the urban aerosol.

Figure 5.1c shows the fraction of the total volume (and hence mass) of the particles in each size range in the same idealised urban aerosol as Figure 5.1a. In this case the curve shows a bimodal distribution and the complete absence of a shoulder to lower particle sizes. The volume (or mass) distribution peaks at particle sizes of about 0.3 and 6 µm, respectively, with total volumes of about 20 and 30 µm³ /cm³ under each peak. The volume (or mass) of the urban aerosol appears to be in two modes: the accumulation mode and the coarse mode, with the latter containing particles in the size range 2 µm and above. The

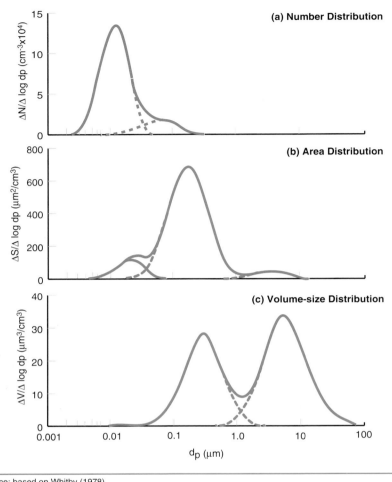

Figure 5.1 Idealised Particle Size Distribution of Urban Aerosols in the USA.

Source: based on Whitby (1978).

nucleation mode particles, although the greatest in number contain a negligible fraction of the aerosol volume and mass.

Most of the particles in the coarse mode are formed by the frictional processes of comminution, such as wind-suspended soil dust or evaporated sea-spray (primary suspended particulate material) and from the slow growth of particles from the accumulation mode (secondary particulate material). Typically, there might be a few tens or hundreds particles per cm³ in the coarse mode in an urban area. However, not all of the coarse mode particles pass through the PM_{10} sampling and monitoring system and are measured. Coarse mode particles account for generally about 20-50% of the urban PM_{10} mass in the UK.

The contributions to the particle size distribution of the different types of atmospheric aerosol are visualised in Figure 5.2, taken from Slinn (1983). Around 1 µm there is a saddle point which separates on the one hand the suspended particulate material formed by storms, the oceans and volcanoes, from the fine particulate matter formed by fires, combustion and chemistry.

Following a consideration of these idealised aerosol distributions, attention is turned to the situation in the United Kingdom and to what is known about the concentrations of secondary particulate material derived from the following sources:

- sulphuric acid and ammonium sulphate in particulate form, derived from the oxidation of sulphur dioxide emitted by combustion sources,

Figure 5.2 Major Sources of Aerosol Particles.

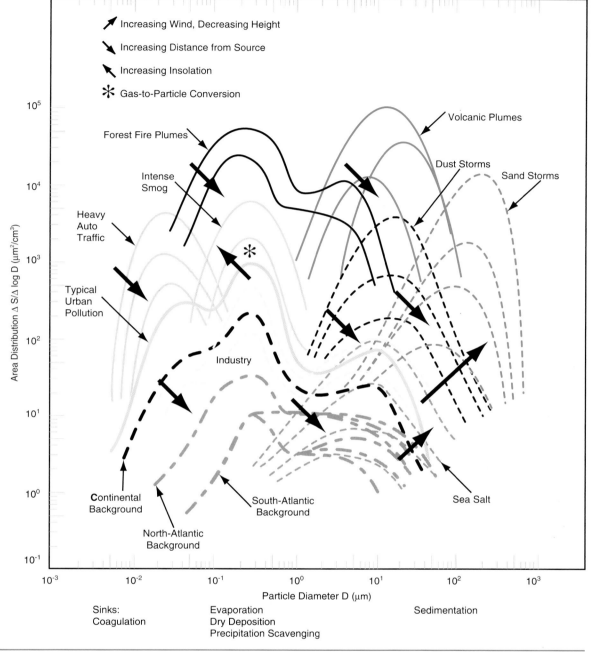

Source: Slinn (1983).

- ammonium and other nitrates in particulate form, derived from the oxidation of nitrogen oxides emitted by combustion sources,

- semi-volatile organic compounds adsorbed onto the atmospheric aerosol, derived from motor vehicles, the resuspension of previously deposited material, industrial processes and stationary fuel combustion,

Clearly, from the definitions adopted here, none of the above sources have been taken into account in producing the emission inventories described elsewhere in this report.

5.2 CONCENTRATIONS OF SECONDARY PARTICULATE MATERIAL IN THE UNITED KINGDOM

5.2.1 Secondary Particulate Material and the Urban Monitoring Network Data

5.2.1.1 Seasonal Variations in Daily Mean PM_{10} Concentrations

Highest daily mean PM_{10} concentrations in each month (see Figure 5.3) show evidence of substantial seasonal cycles at all the national Automatic Urban Network (AUN) urban background sites. These seasonal cycles are characterised by winter maxima as shown by most urban pollutants. Summertime levels are typically lower than wintertime values. However, relative to wintertime levels, summertime levels do not always fall as low as they do for other primary urban pollutants. This behaviour is usually explained by the presence of a significant summertime source in addition to the wintertime primary source. This additional source is the formation of PM_{10} by the photochemical oxidation of sulphur dioxide and nitrogen oxides.

The primary and secondary contributions to urban PM_{10} levels have different sources, characteristics and spatial scales. The primary contribution tends to be combustion- and traffic-derived and is highly localised in the vicinity of its sources. The secondary contribution is largely sulphuric acid, ammonium sulphate and ammonium nitrate, and is much more evenly distributed, showing little variation across an urban and industrial area. Before further progress can be made towards understanding the impact of present policies on PM_{10} air quality and the significance of the rollback required to meet the air quality standard recommended by EPAQS, there is an important requirement to quantify the relative contributions from primary and secondary sources in the different urban areas and with time of year. This is a large task and one for which we have only preliminary results available, at present. These have largely been obtained by examining the correlation between PM_{10} and traffic pollutants at the AUN sites. The traffic pollutants studied are benzene, carbon monoxide, and NO_x, all of

Figure 5.3 Maximum Daily Mean PM_{10} Concentration, 385 Site Months 1992-1994.

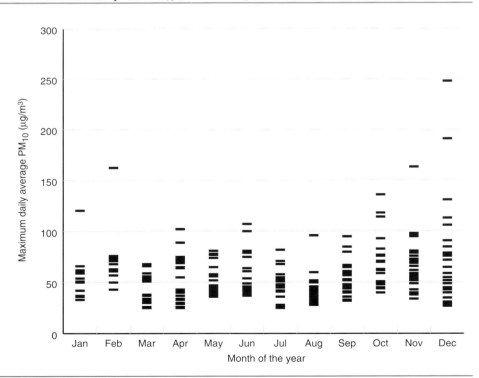

Note: Based on ratified data.

which pollutants have well characterised motor vehicle sources.

5.2.2 Particulate Sulphuric Acid and Ammonium Sulphate

The earliest urban measurements of particulate sulphuric acid were made during the London smog episode in December 1952, when reported concentrations were especially high. Continuous measurements of both particulate sulphuric acid and sulphate aerosol in the United Kingdom were made as part of the Teesside Mist Study in the late 1960's. Although urban concentrations were significantly elevated, on occasions rural concentrations could be elevated as well by the drift of suspended particulate material out of the industrial conurbations.

With the discovery of photochemical ozone formation in the United Kingdom and the associated haze production (Atkins, Cox and Eggleton, 1972), much of the interest in particulate sulphur compounds switched from their health impacts to their role in visibility reduction. Elevated hourly concentrations of particulate sulphuric acid and ammonium sulphate were reported together with ozone concentrations in the range 80-100 ppb, during July 1971, at Harwell, Oxfordshire.

Long term time series of total particulate sulphate concentrations are available from 1954 onwards for Chilton near Harwell in Oxfordshire (Lee and Atkins, 1994) and are shown in Figure 5.4. The concentrations are reported as µg S/m^3 and show a steady rise through the 1950s and 1960s to a maximum in the 1970s, when concentrations started to fall through to the present day. Because of the relatively long lifetime of particulate sulphate, these concentrations are thought to be representative of a rather wide area of southern England.

The rise in particulate sulphate observed at Harwell, Oxfordshire since the 1950s and its subsequent decrease is reflected in the longer term record of sulphur deposition in Greenland ice (Mayewski et al, 1986). Here the growth in use of sulphur-containing fossil fuels can be clearly detected as setting in around the mid-19th century. A more recent decline is also detectable showing that the pattern observed in southern England is likely to have been reproduced throughout much of continental Europe.

The growth in interest in acid rain and environmental acidification in Europe during the 1970s prompted the establishing of the UNECE European Measurement and Evaluation Programme (EMEP) and a monitoring network for particulate sulphate in rural areas across Europe. By the 1990s, a network of over one hundred stations has been established, reporting monthly mean particulate sulphate concentrations (Schaug et al, 1994). Results are presented in Figure 5.5 for the entire EMEP network, showing a clear gradient in levels across the UK from about 1.5 µg S/m^3 in the south and east to about 0.5 µg S/m^3 in the north and west, which approaches a factor of three.

Urban monitoring of particulate sulphate levels started in the UK during the 1970s at a handful of sites and

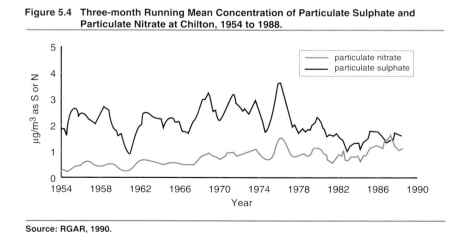

Figure 5.4 Three-month Running Mean Concentration of Particulate Sulphate and Particulate Nitrate at Chilton, 1954 to 1988.

Source: RGAR, 1990.

Figure 5.5 Sulphate in Aerosols Across Europe, 1993.

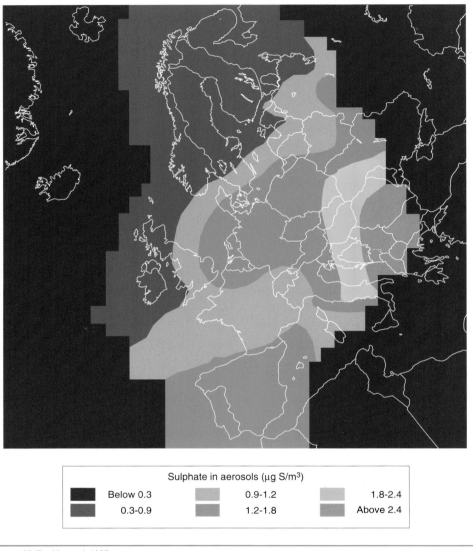

Sulphate in aerosols (µg S/m³)
- Below 0.3
- 0.3-0.9
- 0.9-1.2
- 1.2-1.8
- 1.8-2.4
- Above 2.4

Source: Hjellbrekke et al, 1995.

confirmed the importance of photochemical oxidation as the main source, as in the rural areas. Mean UK urban concentrations tended to be similar to rural levels, confirming that long range transport rather than urban scale production was the dominant source of the elevated urban summertime particulate sulphate concentrations.

With the onset of national urban PM_{10} monitoring in the United Kingdom in 1992, increasing interest has been given to assessing the contribution made by particulate sulphuric acid and ammonium sulphate to urban PM_{10} levels. Almost all the rural measurements of particulate sulphur compounds are reported as µg S/m³ and so are not directly comparable with urban PM_{10} measurements reported as µg/m³. A simple conversion factor does not exist, but a rather good guide can be obtained if results in µg S/m³ are scaled by a factor of about 4. Assuming that rural concentrations can be taken as a guide to urban concentrations, baseline urban levels of particulate sulphate of the order of 6 µg/m³ on a PM_{10} basis can be anticipated.

The strongest evidence that photochemically-generated secondary particulate sulphate contributes to urban PM_{10} measurements comes from an analysis of the AUN PM_{10} data itself. Monthly mean PM_{10} concentrations exhibit a significantly different seasonal cycle to that of the typical primary pollutants such as carbon monoxide, NO_x and some of the VOCs. PM_{10} levels tend to be highest in autumn and winter, as do the typical primary pollutants, but differ in that summer mean levels are only slightly lower, whereas they are significantly lower for the primary pollutants. These points are illustrated in Table 5.1 and Figure 5.3 with the available national PM_{10} data for 1992-1994.

PM_{10} levels tend to rise during summertime photochemical episodes and at several sites simultaneously in response to the changing meteorological situation (see Figure 5.6 and Figure 8.4). In this respect, the behaviour shown by urban PM_{10} concentrations is reminiscent of that of rural ozone concentrations. Summertime maximum hourly ozone concentrations appear to correlate well with maximum daily mean PM_{10} concentrations at some AUN sites. Hourly PM_{10} and ozone concentrations tend not to correlate well because ozone shows a characteristic diurnal variation which is not shared by PM_{10}, since it lacks the efficient nocturnal depletion mechanisms of dry deposition and chemical reaction with NO exhibited by ozone.

The highly time-resolved hourly measurements of particulate sulphuric acid and ammonium sulphate concentrations reported during the photochemical episode during 1971 show that it is indeed likely that much or all of the urban PM_{10} currently measured during the summertime could be accounted for by the regional scale transport of photochemically oxidised sulphur compounds. In which case, the majority of the particulate sulphur is anticipated to be present in the accumulation mode, in the particle size range 0.1-0.3 µm. In this size range its dry deposition velocity is at a minimum, and its lifetime many days.

5.2.3 Particulate Ammonium and Other Nitrates

Although particulate nitrate and sulphate share many common features, their formation mechanisms and atmospheric behaviour are in fact rather different. The processes involved with particulate nitrate are only being slowly understood and are complicated by measurement inadequacies, artefacts and lack of coverage.

Nitrate can be detected in suspended particulate material and its presence usually correlates well with that of sulphate showing some measure of association. The long-term time-series of particulate nitrate concentrations measured at Harwell, Oxfordshire (Lee and Atkins, 1994) shows a steady rise throughout the period from 1954 onwards, without the recent decrease observed for particulate sulphate (Figure 5.4). Again, this behaviour parallels that of nitrate in Greenland ice (Mayewski et al, 1986) since preindustrial times.

The UNECE EMEP programme has established a network for total inorganic nitrate, which includes both particulate nitrate and gaseous nitric acid, although fewer sites have contributed on a historic basis compared with those measuring particulate sulphate. Table 5.2 shows the data for the UK monitoring sites. Clearly human activities produce elevated nitrate aerosol concentrations and a gradient is likely to exist in concentrations across the UK as with sulphate aerosol.

No long-term routine measurements of particulate nitrate aerosol appear to have been made in urban areas of the UK. Measurements have been reported for urban and rural areas of Essex by Harrison and Allen (1990). There is no reason to expect that particulate nitrate aerosol is not an ubiquitous contributor to urban PM_{10} levels. Indeed, measurements made in the Los Angeles basin show that particulate nitrate is the major inorganic contributor to urban PM_{10} levels in that city.

The main fate of the NO_x emitted in urban and rural areas is oxidation to nitric acid vapour by hydroxyl radicals during daytime and by ozone to N_2O_5 at night. Sulphuric acid produced by the oxidation of sulphur

TABLE 5.1 Maximum Daily Mean PM$_{10}$ Concentrations for National Network Stes (µg/m^3).

Site	Year	Jan	Feb	Mar	Apr	May	June	July	Aug	Sep	Oct	Nov	Dec
London Bloomsbury	1992	85	68	51	94	47	75	68	41	66	43	43	79
	1993	66	76	68	75	81	100	54	50	44	48	81	29
	1994	42	75	31	41	57	54	58	41	46	93	58	91
Belfast	1992	-	-	20	30	56	45	28	28	95	136	95	248
	1993	120	61	66	73	78	107	25	42	65	118	98	40
	1994	39	162	38	40	57	49	42	30	66	83	55	191
Edinburgh	1992	-	-	-	-	-	-	-	-	-	32	29	72
	1993	36	38	43	43	41	46	-	37	32	45	66	35
	1994	33	43	25	25	42	38	41	28	25	62	38	49
Newcastle	1992	-	-	37	71	66	72	48	51	67	40	40	71
	1993	42	62	54	65	45	79	63	-	-	38	73	43
	1994	50	63	33	43	52	51	51	37	53	77	58	59
Birmingham Centre	1992	-	-	27	69	77	64	71	60	57	59	34	131
	1993	42	26	56	102	57	79	27	40	66	44	71	27
	1994	33	63	30	30	43	43	46	39	36	71	56	113
Cardiff	1992	-	-	-	-	-	38	36	45	52	48	69	85
	1993	61	52	59	89	65	75	54	-	-	-	62	45
	1994	60	68	59	64	58	81	82	96	80	76	72	43
Leeds	1993	57	62	71	50	42	50	29	52	60	51	96	43
	1994	62	59	33	34	41	46	55	51	60	114	76	77
Bristol	1993	37	63	53	75	57	61	58	44	61	49	81	49
	1994	54	71	34	33	47	42	41	40	40	83	49	52
Liverpool	1993	-	-	-	93	74	-	29	44	85	48	163	49
	1994	51	76	32	29	52	39	47	39	58	84	59	76
Birmingham East	1993	-	-	-	-	-	-	-	-	-	-	-	31
	1994	51	57	26	26	37	41	45	32	33	62	79	106
Southampton	1994	50	39	33	55	39	45	48	35	40	63	53	43
Leicester	1994	59	50	25	30	36	39	42	40	33	51	52	65
Hull	1994	56	73	34	37	58	37	52	46	48	85	62	55
London Bexley	1994	-	-	-	-	27	44	58	33	41	70	58	49
Swansea	1994	-	-	-	-	-	-	-	-	-	-	-	46

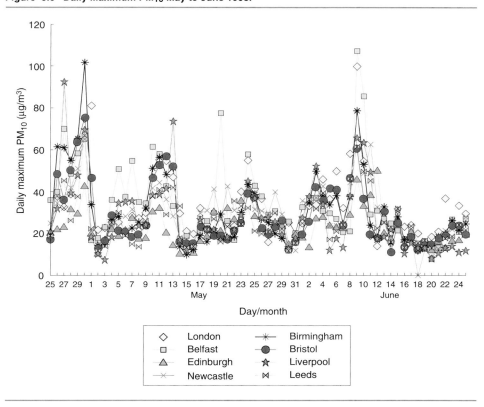

Figure 5.6 Daily Maximum PM_{10} May to June 1993.

Table 5.2 Nitrate Plus Nitric Acid Data from the EMEP Monitoring Network, 1992 (µg N/m³).

Site	Annual Mean	Standard Deviation	Minimum	Maximum
Eskdalemuir	0.39	0.63	0.0	4.93
High Muffles	0.88	0.95	0.03	6.51

Source: Schaug et al, 1994.

dioxide can readily self-nucleate to form particulate material or can be adsorbed to pre-existing particles to produce accumulation mode aerosol. Nitric acid vapour does not share this property of self-nucleation. Furthermore, those accumulation mode particles which are loaded up with acidic sulphur compounds are not expected to take up nitric acid vapour efficiently. Nitric acid can however be adsorbed by sea-salt particles displacing HCl, leaving sodium nitrate aerosol. Alternatively, it may react with gaseous ammonia to form ammonium nitrate which may itself be adsorbed by accumulation mode particles. During nightime, N_2O_5 itself reacts with sea salt particles to produce sodium nitrate aerosol. So there are mechanisms by which particulate nitrate can be formed; however, they are likely to have a markedly variable efficiency compared with those for particulate sulphate.

The evidence that particulate nitrate is associated with summertime photochemical episodes is more difficult to find. Nevertheless, it is likely that regional scale photochemical production of particulate nitrate does occur and that this contributes to urban PM_{10} concentrations in the UK. Almost all the rural measurements of particulate nitrate are reported as µg N/m³ and so are not directly comparable with urban PM_{10} measurements reported as µg/m³. A simple conversion factor does not exist, but a rather good guide can be obtained if results in µg N/m³ are scaled by a factor of about 4. Assuming that rural concentrations can be taken as a guide to urban concentrations, baseline urban levels of particulate nitrate of the order of 4 µg/m³ on a PM_{10} basis can be anticipated. This secondary contribution to urban PM_{10} is somewhat smaller than that of particulate sulphate but is still quite significant.

5.2.4 Impact of Present Policies On Future Secondary PM_{10} Air Quality

The spatial distribution of the secondary aerosol concentrations has been calculated with the Hull Acid Rain Model (HARM) at a spatial resolution of 20kmx20km across the United Kingdom. The model treats both European and UK sources of sulphur dioxide, nitrogen oxides, ammonia and hydrogen chloride, together with their interacting chemistry, dry deposition and wet removal. The spatial patterns calculated of the particulate concentrations of sulphate, nitrate, chloride and ammonium agree with the available observations, such as they are. Figure 5.7 illustrates the geographical variation in the model-calculated concentrations, with a peak in the south east and a minimum in the north west of Britain. The model aims to give a detailed source attribution of the secondary PM_{10} aerosol species across the United Kingdom. Full details of the model are given elsewhere (Metcalfe et al 1995).

It has been possible to estimate how the distribution of secondary PM_{10} will respond to policy measures already in hand by developing an emission scenario for the year 2010. This scenario has then been used in the HARM model to estimate future secondary PM_{10} contributions. The elements of the emission scenario are as follows:

- the Oslo Protocol to the UNECE international convention on Long Range Transboundary Air Pollution, which should lead to about a 60% reduction in sulphur dioxide emissions across Europe by the year 2010 on present day levels,

- the Sofia Protocol to the UNECE international convention which should lead to a freeze on nitrogen oxide emissions at 1987 levels,

- the installation of three-way catalyst technologies on petrol-engined motor vehicles in the EU countries and Norway and some limited action on diesel vehicle emissions during the late 1990s in line with the proposals of the European Commission,

- NO_x emissions from Large Combustion Plant in the UK have been assumed to decrease in accord with the EC Directive and stay constant from 1997 onwards,

- HCl emissions are assumed to decrease in line with power station SO_2 emissions,

- no action has been assumed on ammonia emissions across Europe.

Mean PM_{10} contributions across Great Britain are tabulated in Table 5.3 in both scenarios. There appears to be a 39% reduction in secondary PM_{10} to be anticipated by the year 2010 due to present policies. The result will be a future secondary aerosol which is largely composed of nitrate.

5.2.5 Semi-Volatile Organic Compounds

It has been long recognised that large quantities of organic compounds are emitted into the urban atmosphere. The main sources involve evaporation or fuel spillage, natural gas leakage, solvent usage, industrial and chemical processes and the use of bitumen and road asphalt. The sources of much of the

Table 5.3 Contribution from the Major Aerosol Species to PM_{10} in the Present Day and Future Atmospheres as Calculated in the HARM Model.

Contribution	Present Day PM_{10}, µg/m³	Future 2010 PM_{10}, µg/m³
Sulphate	2.8	1.2
Nitrate	2.9	2.2
Ammonium	0.7	0.5
Chloride	0.1	0.1
Total	6.5	4.0

Source: Metcalfe et al, 1995

Figure 5.7 Secondary PM$_{10}$ Concentration as Calculated by the HARM Model.

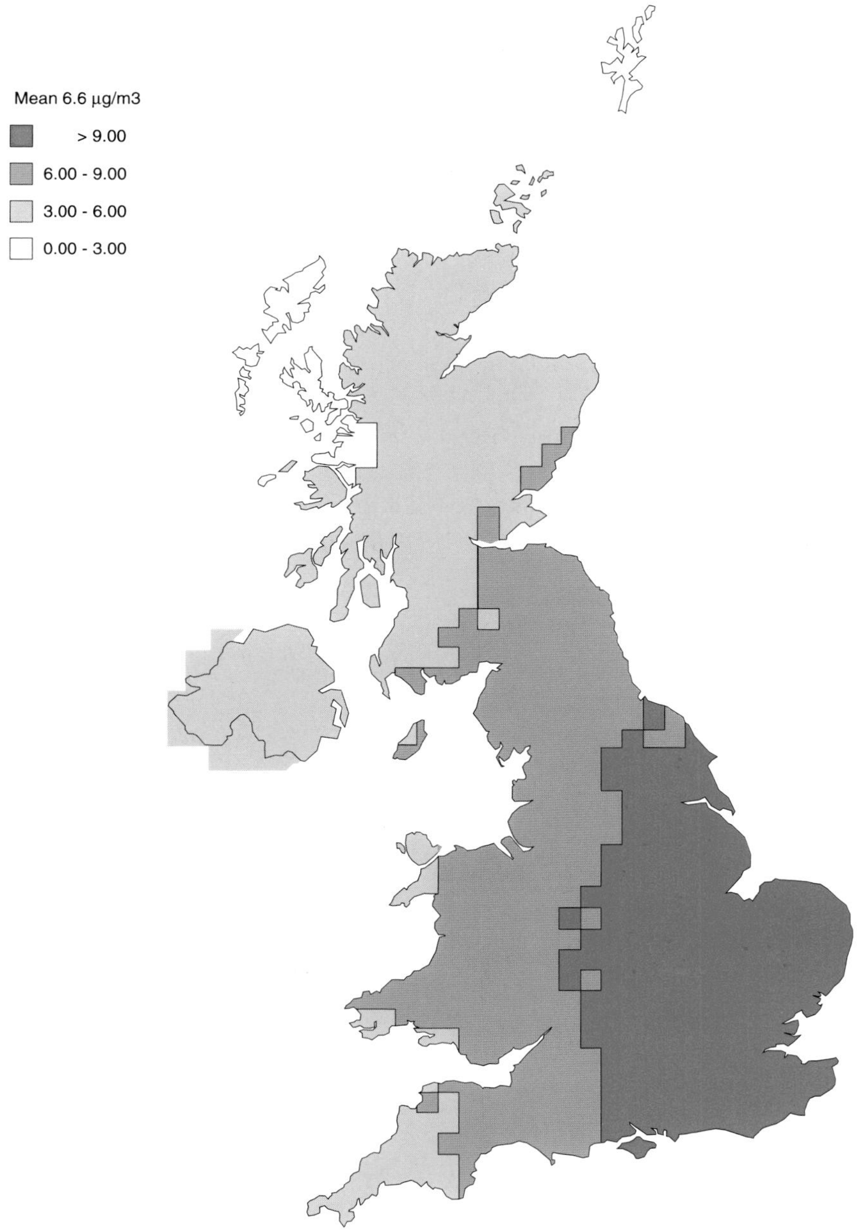

Mean 6.6 µg/m3
- > 9.00
- 6.00 - 9.00
- 3.00 - 6.00
- 0.00 - 3.00

Source: Metcalfe et al, 1995.

low-molecular weight organic compounds have been quantified for the UK. This material is of high volatility generally and is readily removed from the atmosphere over a period of days to years by atmospheric oxidation. However, as molecular weight increases, volatility decreases and there comes a point when the atmospheric behaviour of the more complex organic compounds changes radically from that exhibited by the low-molecular weight compounds.

For some organic compounds volatility is so low that evaporation is an exceedingly slow process. Emissions to the atmosphere occur through some form of high temperature process such as firing, curing or melting as with the laying of road asphalt. Under these circumstances, the emitted organic compounds rapidly cool and condense on any pre-existing particles in the atmosphere. Although nucleation mode particles are the most numerous in urban areas, they do not offer the large superficial area as that provided by the

Table 5.4 Concentrations of Organic Compounds Formed in Air Samples Taken in Belgium, (reprinted courtesy of Elsevier Science Ltd).

Compound	Concentration (ng/m³) Particle Samples P	Gas-phase Samples G	Distribution Factor P/G
Aliphatic Hydrocarbons			
n-nonadecane	0.80	15.1	0.053
n-eicosane	0.85	7.55	0.113
n-heneicosane	1.08	4.12	0.262
n-docosane	2.33	4.23	0.551
n-triosane	4.75	3.38	1.41
n-tetracosane	8.15	4.63	1.76
n-pentacosane	9.50	5.74	1.66
n-hexacosane	9.73	8.70	1.12
n-heptacosane	11.1	9.03	1.39
n-octasane	8.10	7.80	1.04
n-nonacosane	15.8	7.32	2.41
n-tricontane	5.75	4.87	1.18
n-hentriacontane	11.2	3.99	2.81
Polyaromatic Hydrocarbons			
Phenanthrene and anthracene	1.21	44.7	0.027
Methylphenanthrene and methylanthracene	0.90	10.2	0.088
Fluoranthene	2.22	8.52	0.261
Pyrene	3.17	3.36	0.488
Benzofluorenes	2.33	1.87	1.246
Methylpyrene	0.93	-	P
Benz(a)anthracene and chrysene	12.2	3.87	3.15
Benzo(k)fluoranthene and benzo(b)fluoranthene	23.1	2.01	11.5
Benzo(a)pyrene, benzo(e)pyrene and perylene	20.1	2.69	7.47
Phthalic Acid Esters			
di-isobutylphthalate	1.73	32.8	0.053
di-n-butylphthalate	101	353	0.286
di-2-ethylhexylphthalate	54.1	127	0.426
Miscellaneous			
Anthraquinone	1.59	5.66	0.281
Fatty Acid Esters			
Lauric acid	0.01	30.3	0.0003
Myristic acid	1.39	7.58	0.183
Pentadecanoic acid	3.60	5.35	0.673
Palmitic acid	29.0	4.77	6.08
Heptadecanoic acid	2.84	5.71	0.497
Oleic acid	2.06	-	P
Stearic acid	35.7	2.27	15.73
Nonadecanoic acid	1.91	1.02	1.87
Eicosanoic acid	9.04	3.00	3.01
Heneicosanoic acid	2.56	1.65	1.55
Docosanoic acid	13.7	-	P
Tricosanoic acid	3.23	-	P
Tetracosanoic acid	10.7	-	P

Table 5.4 Concentrations of Organic Compounds Formed in Air Samples Taken in Belgium, (reprinted courtesy of Elsevier Science Ltd)(cont.).

Compound	Concentration (ng/m³) Particle Samples P	Gas-phase Samples G	Distribution Factor P/G
Pentacosanoic acid	2.55	-	P
Hexacosanoic acid	9.12	-	P
Aromatic acids			
Pentachlorophenol	2.43	-	P
Basic Compounds			
Acridine, phenanthridine and benzoquinolines	0.94	-	P
Benzacridines	0.85	-	P

Source: Cautreels and Van Cauwenberghe (1978)
Low molecular-weight organics found only in the gas phase were not included in this extract of the authors' original table.
P in the last column indicates that essentially all of the material was in particle form.

accumulation mode particles. So the high-molecular weight organic compounds tend to attach themselves rapidly in timescales of seconds to minutes to the accumulation mode particles and hence contribute to urban PM_{10} levels.

Table 5.4 illustrates how with increasing molecular weight and decreasing volatility, organic compounds become increasingly attached to particles rather than remain in the gaseous phase. This data is taken from Cautreels and Van Cauwenberghe (1978) and refers to measurements made in urban areas of Belgium. The table confirms that a wide range of aliphatic hydrocarbons, polyaromatic hydrocarbons, oxygenates, aromatic acids, esters and nitrogen-containing organic compounds are taken up by the accumulation mode aerosol and contribute to the secondary particulate material present in urban areas.

5.3 KEY POINTS

- *Secondary particles are formed in the atmosphere, mostly from the oxidation of sulphur and nitrogen oxides.*

- *The main components of secondary particulate matter are ammonium nitrate and ammonium sulphate from ammonia neutralisation of sulphuric and nitric acids, and semi-volatile organic matter.*

- *Airborne concentrations of sulphate measured in southern England have decreased in recent years, whilst concentrations of nitrate have increased steadily since measurements began in 1954.*

- *Numerical modelling of the impact of currently agreed controls on emissions of sulphur and nitrogen oxides predicts a reduction in secondary PM_{10} concentrations across the UK of 39% by the year 2010. Nitrate will be the dominant component at that time.*

- *A wide range of semi-volatile organic compounds can condense on atmospheric particles. Few measurements of this component are available in the UK.*

REFERENCES

Atkins DHF, Cox RA and Eggleton AEJ (1972) **Photochemical Ozone and Sulphuric Acid Formation in the Atmosphere over Southern England**, Nature, 235, 372-376.

Cautreels W and Van Cauwenberghe K (1978) **Experiments on the Distribution of Organic Pollutants Between Airborne Particulate Matter and the Corresponding Gas Phase**, Atmos Environ, 12, 1133.

Harrison RM and Allen AG (1990) **Measurements of Atmospheric HNO_3, HCl and Associated Species on a Small Network in Eastern England**, Atmos Environ, 24, 369-376.

Hjellbrekke AG, Lovblad G, Sjoberg K, Schaug J and Skjelmoen JE (1995) **EMEP Data Report 1993**, EMEP.

Mayewski PA, Lyons WB, Spencer MJ, Twickler M, Dansgaard W, Koci B, Davidson CI and Honrath RE (1986) **Sulphate and Nitrate Concentrations from a South Greenland Ice Core**, Science, **232**, 975-977.

Metcalfe SE, Whyatt D and Derwent RG (1995) **A Comparison of Model and Observed Network Estimates of Sulphur Deposition across Great Britain for 1990 and its Likely Source Attribution**,. Q J Royal Met Soc, 121, 1387-1412.

RGAR (1990) **Acid Deposition in the United Kingdom 1986-1988**, RGAR, London.

Slinn WG (1983) **Precipitation Scavenging, Dry Deposition and Resuspension, Proceedings of the Fourth International Conference**, (Pruppacher HR, Semonin RG and Slinn WG, Coords), Elsevier, New York.

Schaug J, Pederson U, Skjelmoen JE and Kvalvagnes I (1994) **EMEP Data Report 1992**, EMEP.

Whitby K (1978) **The Physical Characteristics of Sulphur Aerosols**, Atmos Environ, 12, 135.

6 Concentrations and Trends in Particulate Matter

6.1 INTRODUCTION

Particulate matter, unlike other ambient atmospheric pollutants, tends to be classified by the measurement technique used. Thus, PM_{10}, $PM_{2.5}$, black smoke, strong acid aerosol, total suspended particulate are all terms which are used to characterize the physical or chemical state of airborne particulate matter and which give an indication of the measurement method employed.

Historically, filter methods were first employed to measure the concentrations of particulate matter. The material collected on a filter paper for a specified interval, typically a day, was analysed by weighing or by determining the intensity of the black stain produced (the "black smoke" or "smoke stain" method). Measurements have been made at certain sites which go back to the early part of this century. Although these sampling methods continue to be used and do provide extensive spatial and temporal coverage, automatic analysers such as the tapered element oscillating microbalance (TEOM) have recently been introduced in the United Kingdom and are increasingly being used to give measurements on a finer timescale. There has been a tremendous increase in the amount of data available in the United Kingdom on particulate matter as PM_{10} since the introduction of these instruments and the emphasis of this Chapter will be primarily on these newer measurements.

The general picture gained from the measurements, especially from the smoke and sulphur dioxide networks, is of a dramatic decline in the levels of particulate matter over the course of the last 40 years. The introduction of smoke control areas after the London smog episodes of 1952 and 1962, the shift away from coal burning for space heating and cooking in most parts of the UK (Northern Ireland excepted) and the "tall stack" policy have greatly reduced the emissions of particulate matter and their ambient concentrations.

Although the levels of particulate matter are now significantly lower than those observed historically, the epidemiological studies undertaken recently, most notably in the United States (Dockery et al, 1993; Pope et al, 1995), have once again focussed attention on current levels of particulate matter. This Chapter will review the measurements made by the different monitoring techniques in both urban and rural environments, will seek to identify trends in the concentration of particulate matter, to discern relationships with other pollutants and ultimately to see whether the sources of particulate matter can be identified.

6.2 MEASUREMENTS OF PARTICULATE MATTER

The different measurement techniques used to characterize particulate matter were described in Chapter 3 of this report. At the present time, measurements of particulate matter are made on a routine basis in three national networks:

(1) the Basic Urban Network for smoke and sulphur dioxide, in which daily concentrations of particulate matter are determined using the "smoke stain" method. The measurements made in this network represent the longest running series of data. Sites were chosen to be representative of locations in major population centres whilst ensuring spatial coverage of the whole country. The network currently comprises 154 sites, 84 of which are also included in the EC Directive network described in (2) below.

(2) the EC Directive network for smoke and sulphur dioxide, in which daily concentrations of particulate matter are again determined using the smoke stain method. This network was established to monitor compliance with the EC Directive (80/779/EC) on Sulphur Dioxide and Suspended Particulate Matter. It currently comprises 155 sites. The measurements are specifically made at sites which were at risk of breaching the Directive or where breaches have occured. The number of sites has fallen from a peak of 330 sites in 1986 as levels of smoke have declined (QUARG, 1993a).

(3) the Automatic Urban Network, in which hourly concentrations of particulate matter are determined as PM_{10} using tapered element oscillating microbalance (TEOM) instruments. The network was established in 1992 and now comprises 16 sites positioned in urban background locations.

- Belfast
- Bristol
- Birmingham (Centre)
- Birmingham (East)
- Cardiff
- Edinburgh
- Kingston-upon-Hull
- Leeds
- Leicester
- Liverpool
- London (Bexley)
- London (Bloomsbury)
- Middlesbrough
- Newcastle
- Southampton
- Swansea

The network is set to expand, with the establishment of a further 13 sites by the end of 1996.

In addition the Department of the Environment has announced that it intends to integrate around 35 locally operated sites into the national network by 1996. Some of these sites are expected to monitor PM_{10}.

Figure 6.1 Basic Urban Network Sites, 1993/94.

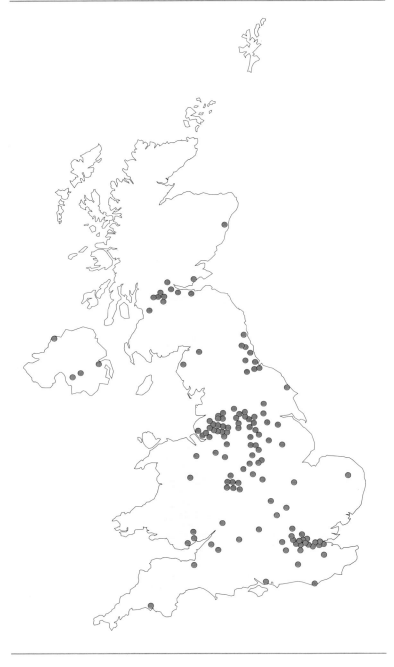

Table 6.2. An Analysis of the Correlations between Hourly PM_{10} Levels Measured at the AUN site in Edinburgh and the Weather Data for Edinburgh Airport for all Months in 1994.

Month	Linear Regression Slopes			
	Windspeed (knots)	Temperature (Degrees C)	Rain during the last hour, 0.1 mm	Sunshine Intensity (W/m²)
January	-0.403*	-0.789*	-0.099	0.0054
February	-0.406	-0.411	-0.114	0.0174
March	0.010	0.346	-0.289*	0.0169*
April	-0.011	0.256	0.036	0.0111*
May	0.566*	0.954*	0.222	0.0070
June	0.281	0.931*	0.110	0.0200*
July	-0.517	1.005*	0.221**	0.0078
August	0.160	0.852*	0.201	0.0168*
September	-0.278	0.588	-0.054	0.0050
October	-0.724*	0.810	-0.111	0.0035
November	-0.432	-0.555	-0.329*	0.0013
December	-0.942**	-0.766*	-0.104	0.0027

Notes:

* indicates a regression correlation coefficient, R > 0.2, n<744.

** indicates R>0.4, n<744.

Based on provisional data

Cardiff, where the monitoring site was affected by building development in the vicinity between March and October 1994 and at the Newcastle site which was unduly influenced by local sources and was re-located in June 1994 to give more representative measurements of an urban background site.

As shown in Table 6.1, the annual mean concentrations of PM_{10} are relatively uniform across the country, lying between 20-34 µg/m³. From these data, there is a slight tendency for mean concentrations to be larger in regions with large traffic densities (eg London) or significant domestic coal burning (eg Belfast). The mean NO_x (nitrogen oxides) concentrations show a similar variation at these sites and the PM_{10}/NO_x ratio is relatively constant except at Belfast and Cardiff where the PM_{10} concentrations are relatively large. (See also Chapter 4).

In comparison with many of the influences described in the sections above, the influences of weather conditions tend to be more subtle. The passage of large scale weather systems causes day-by-day variations in most pollutants, PM_{10} included. If these weather systems are large enough and intense enough, then PM_{10} concentrations at several sites can apparently change simultaneously or with time lags of a few hours or less. Such events tend to be infrequent though and much of the variation in urban PM_{10} concentrations is driven by local variations in emissions and strong correlations are found with other urban pollutants.

A detailed study has been conducted of the correlations between the hourly PM_{10} concentrations measured at the AUN site in Edinburgh and the weather data for Edinburgh Airport for 1994. The analysis is summarised in Table 6.2, for the following hourly weather parameters:

- windspeed in knots,

- temperature in °C,

- rain in the last hour in 0.1 mm,

- sunshine intensity in W/m².

The analysis clearly shows that the correlations between PM_{10} and weather parameters tend to be weak and intensely seasonal. In the wintertime, PM_{10} concentrations appear inversely related to the windspeed. The greater the windspeed the lower the urban PM_{10} concentrations due to the effects of dilution and ventilation. In the summertime, this trend reverses and PM_{10} concentrations tend to increase with windspeed. There are two likely explanations for this phenomenon. The first is that wind-driven resuspension of surface dusts is greater in the dryer summer months. Secondly, at this coastal location, brisk sea breezes are common in the warmer hours of the day when traffic emissions of PM_{10} are highest..

Correlations with temperature show a similar seasonal turn-over from a wintertime inverse relationship to a summertime dependence. Winter days with strong PM_{10} concentration build-up tend to be those with colder mornings. During the summertime, PM_{10} concentrations build up during the day due to traffic activity and secondary particle formation takes place as temperatures rise giving an apparent association.

Rainfall could in principle exert a variety of impacts on PM_{10} concentrations. However, the Edinburgh data show that PM_{10} is inversely correlated with rainfall in the last hour during wintertime and positively correlated during summertime. In the wintertime, rainfall is associated with the passage of frontal systems with increased windspeeds and the general conditions under which urban pollutant build-up is not anticipated. Hence there appears to be a negative correlation between rain in the last hour and PM_{10} levels. In the summertime, rain tends to be more infrequent, more intense and has the potential to resuspend more material from the earth's surface when it falls.

Correlations of PM_{10} concentrations with sunshine intensity examine the relative diurnal profiles of PM_{10} and sunshine. As the mean diurnal curve of PM_{10} adjusts with season to reflect the changing balance of primary and secondary sources, then its correlation with sunshine hours alters. It is best in summer months when secondary paricles play a major role. This effect is illustrated in Table 6.2.

The cumulative distribution frequency of the PM_{10} concentrations observed at four of the sites operational in 1994 indicate that although there appears to be a degree of uniformity in the mean PM_{10} concentrations, differences are apparent when higher percentiles are considered as shown in Figures 6.10a to 6.10d. This reflects the impact that local sources or conditions can have on PM_{10} concentrations.

Further analysis and discussion on the PM_{10} measurements is given in subsequent sections of this and other Chapters in this report.

6.3.3 $PM_{2.5}$ Measurements

Although particulate matter is measured as PM_{10} at the national automatic monitoring sites, there is also an on-going measurement programme being undertaken at Hodge Hill in Birmingham by the City Council. The concentrations of $PM_{2.5}$, PM_{10} and black smoke are simultaneously being determined using two TEOM instruments with appropriate inlets and the smoke stain method respectively. The measurement site is about 70 metres south of an elevated section of the M6 motorway and the measurements have been made since October 1994. The data for the period from January to June 1995 indicate that the daily averages of $PM_{2.5}$, PM_{10} and black smoke lie between 2-60 µg/m³, 4-76 µg/m³ and 1-41 µg/m³ respectively, as shown in Table 6.3 (Appleby, 1995). The concentrations of $PM_{2.5}$ are always lower than those of PM_{10}, varying between 28 to 100% with an average value of 60%. The variation appears to be weather dependent. (See also Chapter 8).

6.3.4 Strong Acid Aerosol Measurements

Strong acid aerosol was measured on a daily basis in London between 1963 and 1972 by the MRC Air Pollution Unit based at St Bartholomew's Hospital. Earlier measurements (from 1957) had only been made during pollution episodes. Samples were collected on filter paper and the total acidity determined by back titration following the addition of sodium tetraborate. The acidity was assumed to be solely due to sulphuric acid. Table 6.4 gives a summary of the measurements made during this period (MAAPE, 1992). The highest daily mean level of H_2SO_4 recorded was 134.1 µg/m³ in 1964 which can be compared with the highest daily and hourly levels of 347 and 678 µg/m³ respectively which were recorded during the December 1962 episode. The

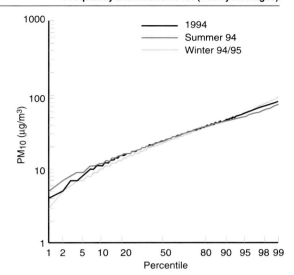

Figure 6.10a London Bloomsbury PM$_{10}$ Cumulative Frequency Distribution Plot (Hourly Averages).

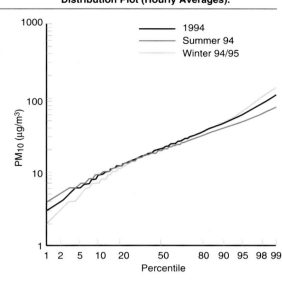

Figure 6.10b Belfast Centre PM$_{10}$ Cumulative Frequency Distribution Plot (Hourly Averages).

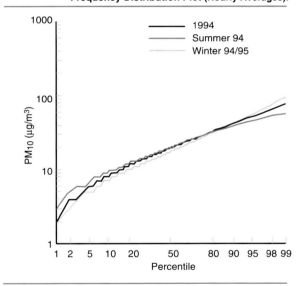

Figure 6.10c Birmingham Centre PM$_{10}$ Cumulative Frequency Distribution Plot (Hourly Averages).

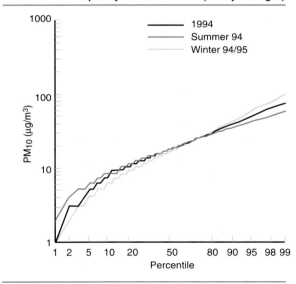

Figure 6.10d Birmingham East PM$_{10}$ Cumulative Frequency Distribution Plot (Hourly Averages).

table shows that between 1964 and 1972 H_2SO_4 concentrations fell by more than 50%.

Sulphuric acid concentrations showed a smaller winter to summer ratio than those observed for sulphur dioxide and other primary pollutants. The ratio indicated that there was a significant source in the summer and two distinct production mechanisms were identified. The photochemical nature of one of the production mechanisms was clearly illustrated by the strong correlation observed between total particulate sulphate and ozone measurements made in central London during a photochemical episode in June 1989. (MAAPE, 1992).

Concentrations of sulphuric acid tend to be lower in rural areas but are harder to interpret reliably because of the interference from ammonia to form ammonium sulphate either in the atmosphere or during sampling and subsequent analysis. One of the few direct

Table 6.3 Comparison of Monthly Average Measurements of Black Smoke, $PM_{2.5}$ and PM_{10} made at Birmingham, Hodge Hill for January to June 1995 (Appleby, 1995).

Month	Average Monthly Value			Ratio*		
	Smoke ($\mu g/m^3$)	$PM_{2.5}$ ($\mu g/m^3$)	PM_{10} ($\mu g/m^3$)	Smoke:$PM_{2.5}$	Smoke:PM_{10}	$PM_{2.5}$:PM_{10}
January	9	11	15	0.80	0.56	0.70
February	4	11	17	0.31	0.21	0.64
March	7	12	22	0.48	0.29	0.60
April	5	15	28	0.34	0.19	0.57
May	6	15	26	0.38	0.20	0.56
June	6	11	24	0.59	0.27	0.48
Minimum Daily Value	1	3	4	0.07	0.04	0.3
Maximum Daily Value	41	43	66	1.72	1.45	1.0
Mean Value	6	13	22	0.48	0.29	0.60

* Mean value of ratios determined for each day in that month.

Table 6.4 Historical London Daily Acid Aerosol Concentration ($\mu g/m^3$ of total H^+ as H_2SO_4) (MAAPE, 1992).

Year	Annual			Winter*		
	Mean	S.D.	Max.	Mean	S.D.	Max.
1964	6.8	9.6	134.1	10.4	15.5	134.1
1965	5.1	3.4	26.3	7.3	5.0	42.2
1966	4.6	3.2	24.3	7.5	4.1	24.3
1967	4.1	2.9	20.2	5.2	2.7	22.5
1968	4.5	2.9	25.3	7.6	3.9	25.3
1969	4.3	3.0	19.0	6.0	3.2	17.0
1970	3.0	2.5	18.2	4.8	2.9	19.0
1971	3.2	2.7	29.7	3.3	3.7	29.7
1972	3.2	2.7	22.9	4.5	2.4	13.7

(*) The winter period is from 1st November of the preceding year to 28th or 29th February of the year shown.

measurements was when simultaneous measurements were made at 3 sites in Essex over 33 days between 1987 to 1989 (Kitto and Harrison, 1992). The measurements gave mean H^+ ion concentrations between 11.5 to 28.7 neq/m^3, equivalent to 0.6-1.4 $\mu g/m^3$ of H_2SO_4. Approximate H^+ ion concentrations have also been inferred from the extended series of measurements of sulphate, nitrate and ammonium made at Harwell assuming ionic balance (i.e. $H^+ = (SO_4^{2-} + NO_3^-) - NH_4^+$ in equivalence units) (MAAPE, 1992; Lee and Derwent, 1995). A simple regression analysis suggested a downward trend in the H_2SO_4 concentration with absolute concentrations in 1991 between 3-5 $\mu g/m^3$. The Harwell measurements are discussed further in section 6.3.6. Although the concentrations observed in Essex were lower than those inferred from the Harwell dataset, this is consistent with the higher NH_3 concentrations expected in Essex.

6.3.5 Other Urban Measurements of Particulate Matter

As part of the London Air Quality Network run on behalf of London Boroughs by the South East Institute of Public Health, TEOM instruments have been set up to measure PM_{10} concentrations at a number of sites across London: Bexley, Greenwich, Haringey, Kensington and Chelsea, Sutton, Tower Hamlets and Thurrock. The site at Bexley is an affiliated site in the national urban network although particulate matter was measured as total suspended particulate prior to affiliation. Data for the sites operational in 1994 (at Bexley, Greenwich and Tower

Chapter 6 : Concentrations and Trends in Particulate Matter

Table 6.5 1994 Statistics for PM_{10} from the Sites in the London Air Quality Network.

Site	Bexley	Greenwich	Tower Hamlets
Site Type	SB	SB	UB
Number of Days Operational	197	321	340
Number of Days on which daily mean levels of PM_{10} (µg/m³) exceeded			
10	188	302	329
20	116	144	194
30	59	51	96
40	33	18	47
50	8	4	12
60	2	4	5
70	1	2	4
80	0	1	1
90	0	0	0
100	0	0	0
Annual Mean Concentration (µg/m³)	25	21	26

Notes: SB - suburban background; UB - urban background. Data from Greenwich and Tower Hamlets are provisional

Table 6.6 Provisional Statistics for PM_{10} from the Roadside Sites in the London Air Quality Network and the National Automatic Urban Monitoring Sites in London for January to June 1995.

Site	London Bloomsbury	London Bexley	Haringey	Thurrock
Site Type	UB	SB	R	UB
Average PM_{10} Concentration (µg/m³)	26	22	22	24
Maximum Hourly PM_{10} Concentration (µg/m³)	249	173	130	108
Number of Days on which Daily Mean Level of $PM_{10} \geq 50$ µg/m³	9	8	6	5
Data Capture %	96	97	91	51

Notes: (1) Site types: R - roadside; SB - suburban background; UB - urban background.
(2) The Thurrock site became operational in March 1995.

Hamlets) are summarised in Table 6.5. The annual mean concentrations of 25, 21 and 26 µg/m³ recorded respectively at the three sites are comparable to the PM_{10} concentrations observed at the urban background sites in the national urban monitoring network.

Two of the sites in the network at Haringey and Sutton are situated in roadside locations. The measurements made at these sites and also at Thurrock for the period from January to June 1995 are presented in Figure 6.11 and are compared in the same Figure with the measurements made at the national sites at London Bexley and Bloomsbury over the same period. The mean and maximum hourly concentrations recorded at certain of these sites between January and June 1995 are given in Table 6.6. The Sutton site only became operational in June 1995. The concentrations observed at the roadside sites are thus far comparable to those measured at the urban background sites during this period.

Other roadside measurements have clearly demonstrated an elevation in PM_{10} concentration close to road traffic. Figure 6.12 shows a time series of PM_{10} measurements made by Westminster City Council at Marylebone Road and Oxford Street in

Figure 6.11 Roadside PM$_{10}$ Data from London Compared to Urban Background PM$_{10}$ Measurements.

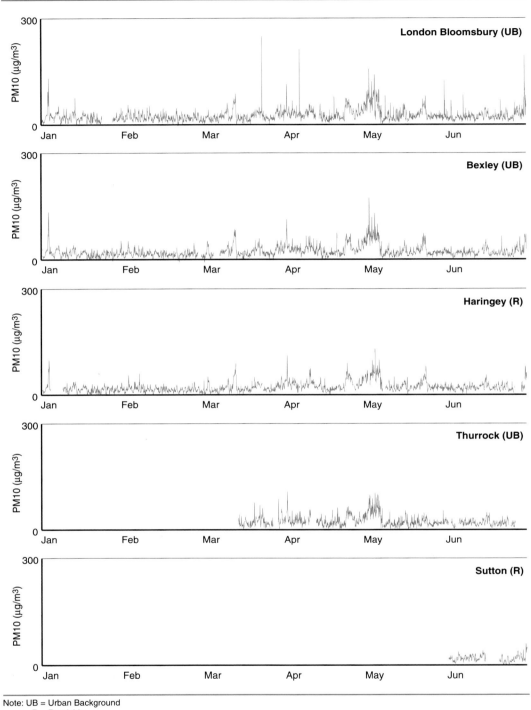

Note: UB = Urban Background
R = Roadside
Data for roadside stations provided by SEIPH
Data for London Bloomsbury and Bexley are provisional.

Central London using TEOM instruments, compared with the AUN site at London, Bloomsbury. The highest concentration over the period monitored was at the Oxford Street site (mean = 52 µg/m³), compared with 36 µg/m³ at Marylebone Road and 29 µg/m³ at Bloomsbury. Oxford Street is notable for being restricted to buses and taxis only. Marylebone Road appears similar to Bloomsbury for the period 21st July to 1st September, when traffic is reduced due to summer holidays. In September and October it shows

Chapter 6 : Concentrations and Trends in Particulate Matter

Figure 6.12 Hourly-mean Concentrations of PM_{10} at Roadside and Urban Background Sites in London, July to October 1995.

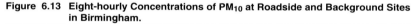

Source: Westminster City Council, 1995.

Figure 6.13 Eight-hourly Concentrations of PM_{10} at Roadside and Background Sites in Birmingham.

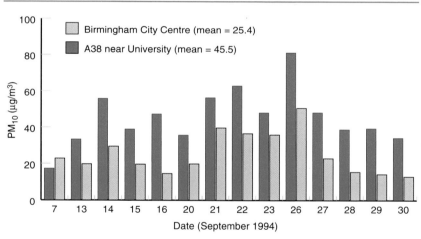

Source: Jones and Harrison, 1995.

a clear elevation, typically by 10-20 µg/m³ above concentrations at the Bloomsbury site.

Eight-hourly daytime measurements of PM_{10} have also been made on the busy A38 road in Birmingham, and compared with data collected simultaneously from the urban background AUN site at Birmingham, Centre. The data (Figure 6.13) show a clear elevation at the roadside site (mean = 45.5 µg/m³) over the background site (mean = 25.4 µg/m³) consistent with that observed for Marylebone Road and Oxford Street in London.

The former London Scientific Services measured concentrations of airborne particulate matter at four sites in London using M-type samplers, with subsequent gravimetric analysis of the particulate matter (QUARG, 1993a). This sampler has a particle size cut-off of approximately 10 to 15 µm (see Chapter 3) and, depending on the wind speed may

Figure 6.14 Annual Mean Concentrations of Suspended Particles at Four Sites in London (1986-1989).

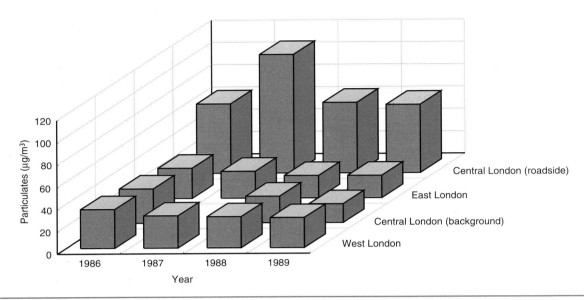

Source: LSS, 1989.

Figure 6.15 Trends in Weekly-mean Suspended Particle Concentrations, Manchester 1980-93.

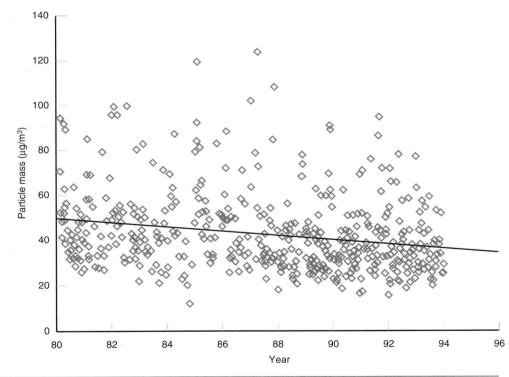

Source: Miller and Lewis, 1996.

therefore tend to under or overestimate PM_{10} concentrations. The samplers were located at three background and one roadside site. The measurements made are summarised in Figure 6.14. The levels measured at the roadside were 2.5 to 3 times higher than those at the background locations, indicating the importance of road traffic as a source of particulate matter. The apparent downward trend for the background locations was believed to be real as other pollutants measured at these sites did not show any significant trend over the period.

An extended series of measurements of suspended particulate matter using M-type samplers was made at Piccadilly Gardens in central Manchester between 1980 and 1993 (Miller and Lewis, 1996). The measurements showed a very strong seasonal behaviour. The dataset was corrected for the seasonal effect and an underlying decline in particulate matter of about 1 µg/m³ per year was observed (Figure 6.15). During 1987 and 1988, a series of measurements was made using an 8-stage Andersen cascade impactor. A pre-separator was used to remove particles with sizes greater than 13.5 µm. The highest concentration was observed for the size range of less than 2 µm with a median value of 1.2 µm. Unfortunately these measurements were not repeated to identify trends in the size distribution.

The national multi-element survey has also shown that the concentration of a range of elements, averaged over the 5 monitoring sites, has fallen considerably over the 17 years of operation of the survey (Loader, 1994).

6.3.6   Rural Measurements of Particulate Matter

Black smoke measurements were made at a number of rural locations as part of the smoke and sulphur dioxide network until 1991. Figure 6.16 shows the annual mean concentrations of black smoke recorded at 4 sites in the network: Caenby, Husborne Crawley, Ratcliffe and Camborne. Although the decline in smoke levels was not as dramatic as those observed at urban sites, there has still been a fivefold reduction between the early 1960s and 1991. The annual mean concentrations recorded in 1990 at the Caenby, Husborne Crawley, Ratcliffe and Camborne sites were 7, 8, 11 and 3 µg/m³ respectively. The values at the urban sites in Belfast, Stoke-on-Trent, Lambeth and Norwich were 17, 19, 25 and 12 µg/m³ respectively. The ratio of urban to rural smoke levels has fallen substantially over this period so that urban levels are now closer to those recorded at rural sites.

A campaign to measure particulate matter concentrations in rural areas was undertaken by TBV Science over a three month period between July and September 1994. Sampling was carried out at 6 sites in the rural West Midlands area using M-type samplers. The average concentrations recorded over this period at the different sites were found to lie between 11 and 20 µg/m³.

A study has recently been undertaken to compare concentrations of PM_{10} at a rural site near Bristol, Chew Lake, with those observed at the national monitoring sites at Bristol and Cardiff (International Mining Consultants Limited, 1995). The monitoring site was chosen to be representative of rural background concentrations and was therefore sited away from any major industrial or local pollution sources. PM_{10} measurements were made using a TEOM instrument for one month between 24 January and 21 February 1995. Figure 6.17 shows the 30 minute mean PM_{10} measurements made at Chew Lake, Bristol and Cardiff during this period. Although there are periods when concentrations of PM_{10} were elevated at the urban sites, these are not apparent in the rural measurements. The rural concentrations were generally low primarily as a result of the wind blowing mainly from the west during the measurement period. The mean concentration of PM_{10} measured during this period was 14 µg/m³ at Chew Lake, 20 µg/m³ for the Bristol site and 23 µg/m³ at the Cardiff site. Further analysis of the measurements led the authors of the study to conclude that rural particulate matter consists of several different components and ashing of representative samples indicated that the component attributable to vehicle exhaust was no greater than 30%. These conclusions are tentative given the limited monitoring period of the study.

There are various other networks from which data on particular chemical constituents in particulate matter (e.g. Na^+, Ca^{2+}, Mg^{2+}, NH_4^+, Cl^-, SO_4^{2-} and NO_3^-) are available. These data can be used to infer the total particle mass. None of these measurements have used inlets designed to sample a defined size fraction. However, like the M-type sampler, the methods used tend not to sample large particles efficiently.

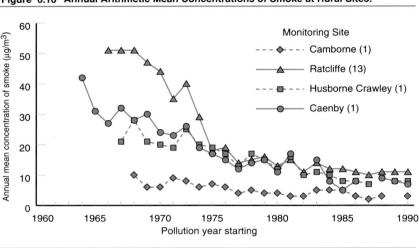

Figure 6.16 Annual Arithmetic Mean Concentrations of Smoke at Rural Sites.

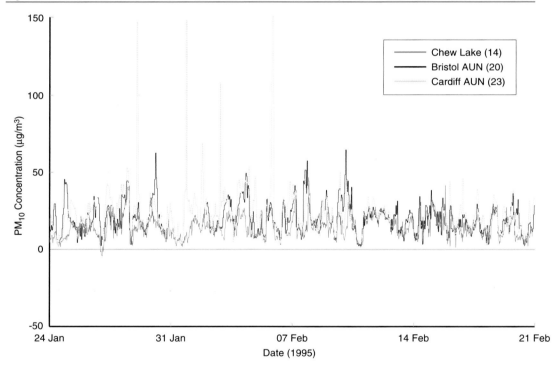

Figure 6.17 30 Minute Mean PM_{10} Concentrations at Chew Lake compared with Bristol and Cardiff.

Note: Data for Bristol and Cardiff are provisional.

At Chilton (SU 472854) in Oxfordshire, measurements of particulate sulphate, nitrate and ammonium have been carried out using a filter sampler since 1954 as shown in Figure 6.18 (Lee and Derwent, 1995). Between 1954 and 1975 the mean concentration of particulate sulphate increased from around 6 µg/m³ to 8 µg/m³ while that of particulate nitrate increased from 1.5 µg/m³ to 4.5 µg/m³. By the 1990s, the mean concentration of particulate sulphate had decreased to 4 µg/m³ and that of particulate nitrate had increased to 5 µg/m³. The total mass of ammonium nitrate and ammonium sulphate present at this site can be calculated if it assumed that the counter cation for the nitrate and sulphate is only ammonium. This suggests that the present concentration is approximately 12 µg/m³ which has fallen from a value of 16 µg/m³ in the mid 1970s.

Table 6.13 Daily Black Smoke Concentrations (μg/m3) Recorded in London, December 1991.

Day	Acton 15	Camberwell 4	City 16	Croydon 15	Ealing 7	Enfield 14	Ilford 6	Islington 9	Kensington 13	Stepney 5	Westminster 17	Mean	Max
1													
2													
3	11	12	18	13	17	8	10	7	14	7	15	12	18
4	11	12	20	16	16	7	10	11	12	9	26	14	26
5	14	14	20	24	21	4	7	11	25	7	17	15	25
6	30	35	35	16	42	10	28	11	49	21	49	30	49
7	60	40	47	14	71	16	61	7	41	26	68	41	71
8	45	54	53	13	63	11	44	4	54	24	65	39	65
9	45	61	62	11	63	11	52	16	55	24	48	41	63
10	32	31	42	8	42	11	1	11	43	20	40	26	43
11	57	58	63	13	82	22	65	12	49	26	52	45	82
12	153	125	140	15	209	24	228	7	166	31	148	113	228
13	162	80	128	15	167	26	181	12	158	36	121	99	181
14	110	88	105	10	124	13	80	12	145	36	99	75	145
15	81	38	81	13	71	17	58	7	84	31	81	51	84
16	44	25	37	25	35	7	24	7	31	24	38	27	44
17	41	8	15	12	15	13	14	4	16	8	14	15	41
18	12	5	14	13	8	2		3	9	6	7	8	14
19	11	4	19	26	4		14	7	5	6	5	10	26
20	9	6	9	34	5		7	11	7	8		11	34
21	2	2	5	16	3		6	7	7	5		6	16
22	3	2	8	28	1		4	7	4			7	28
23	7	4	14	8	7		6	11	3		5	7	14
24	16	15	15	32	17	4	18	11	12		17	17	32
25	14	12	20	13	12	3	13	3	12		12	12	20
26	4	4	10	13	3	3	3	7	4		3	6	13
27	23	28	15	11	23	8	28	7	21		26	19	28
28	27	12		16	25	4	32	3	21			17	32
29	21	19		22	25	3	15	7	14			16	25
30	34	20		20	38	8	24	7	19			21	38
31													
Mean	39	29	40	17	43	11	38	8	38	19	43	30	
Max	162	125	140	34	209	26	228	16	166	36	160		

Table 6.14 Limit and Guide Values for Smoke and Sulphur Dioxide (in µg/m^3) Specified in EC Directive on Smoke and Sulphur Dioxide (80/779/EC).

(1) Limit Values

Reference Period	Limit Values for Smoke	Limit Values for Sulphur Dioxide
Pollution Year (median of daily values)	68	120 if smoke $\leq$ 34 80 if smoke > 34
Pollution Year (98th percentile of daily values)	213	350 if smoke $\leq$ 128 250 if smoke > 128
Winter (median of daily values Oct-Mar)	111	180 if smoke $\leq$ 51 130 if smoke > 51

(2) Guide Values

Reference Period	Guide Values for Smoke	Guide Values for Sulphur Dioxide
Year (arithmetic mean of daily values)	34-51	40-60
24 hours (daily mean values)	85-128	100-150

Note: Limit and Guide Values given for smoke are for the BS calibration as used in the UK. The limits stated in the EC Directive relate to the OECD method; the relationship between concentration determined using the OECD and BS calibration is: BS = 0.85 OECD.

1992, the best estimate of the mean PM_{10} increase across London during the episode is 110 µg/m³. On this basis, the increase in black smoke and PM_{10} could explain some of the increase in mortality and morbidity observed in London (Anderson et al, 1995).

6.6 EXCEEDENCES OF AIR QUALITY STANDARDS AND GUIDELINES

Air Quality guidelines and standards currently exist for particulate matter characterized as black smoke or total suspended particulate and PM_{10}. The EC Directive (80/779/EC) on Sulphur Dioxide and Suspended Particulate Matter gives separate guide and limit values for smoke and suspended particulate matter. As stated earlier, there is no simple relationship between black smoke and total suspended particulate and the United Kingdom has chosen to implement the Directive using black smoke measurements (Annex 1 of the Directive) (Table 6.14). The Directive also gives guide and limit values for sulphur dioxide but these are dependent on the associated concentration of smoke (Table 6.14). This linkage of the two pollutants arose from the observations and epidemiological studies of the London smog episodes. The limit and guide values for suspended particulate matter are currently used in the Department of Transport's Design Manual for Roads and Bridges (DOT, 1994).

Table 6.15 gives the number of exceedences of the EC Directive from the 1987/88 pollution year to the 1994/95 pollution year. In the pollution year 1993/94, there were no exceedences of the EC Directive. As part of the Directive, a number of derogation areas were established, principally where coal remained a significant fuel source and breaches of the Directive could possibly occur. Since 1993, such derogations no longer exist.

The World Health Organization has recommended air quality guidelines for particulate matter, which are linked to the concentration of sulphur dioxide. The current guideline values are summarised in Table 6.16

Table 6.15 Exceedences of the EC Directive for SO_2 and Suspended Particulate Matter.

Site	Pollution Year/Criterion							
	1987/88	1988/89	1989/90	1990/91	1991/92	1992/93	1993/94 (a)	1994/95 (a)
Crewe 17	Sm≥3cd	-	-	-	-	-	-	-
Mexborough 19	Sm≥3cd	-	-	-	-	-	-	-
Seaham 2	Sm/98, Sm≥3cd	-	-	-	-	-	-	-
Hetton-le-Hole 3	Sm/98, Sm≥3cd	Sm/98	Sm/98, Sm>3cd	-	-	-	-	-
Houghton-le-Spring 3	Sm/98, Sm≥3cd	-	-	-	-	-	-	-
Sunderland 8	Sm/98	-	-	-	-	-	-	-
Featherstone 1	Sm/98, Sm≥3cd	-	-	-	-	-	-	-
Washington 4	Sm≥3cd	-	-	-	-	-	-	-
New Ollerton 2	-	Sm≥3cd	-	-	-	-	-	-
Durham Sherburn 1	-	-	-	-	Sm/98	-	-	-
Grimethorpe 3	-	-	-	-	-	(b)	-	-
Belfast 11 (c)	-	-	-	Sm/98, Sm≥3cd	-	-	-	-
Belfast 12 (c)	-	-	-	Sm/98, Sm≥3cd	-	-	-	-
Belfast 39 (c)	Sm≥3cd	Sm/98, Sm≥3cd	-	-	-	-	-	-
Belfast 42 (c)	-	-	Comb/98	Comb/98 Comb≥3cd	Comb/98	-	-	-
Newry 3 (c)	-	-	Sm/98, Sm≥3cd	-	-	-	-	-

Criteria (1) Sm/98 - limit value for the 98th percentile of daily values in a year exceeded for black smoke.

        (2) Sm≥3cd - at least one period of 3 or more consecutive days when daily values were above the 98th percentile limit value for black smoke.

        (3) Comb/98 - both black smoke and SO_2 98th percentiles exceeded the combined limit values (i.e. 128 µg/m³ BS or 150 µg/m³ OECD for smoke and 250 µg/m³ for SO_2).

        (4) Comb≥3cd - at least one period of 3 or more consecutive days when daily values were above the 98th percentile limit values of both black smoke and sulphur dioxide.

Notes (a) There were no exceedences of the Directive in 1993/94 and 1994/95.

       (b) The only exceedence in 1992/93 was at the Grimethorpe 3 site when the limit value for the 50th percentile of daily values in a year was exceeded for SO_2.

       (c) Prior to 1993, a number of derogation areas existed, one of which covered the Belfast sites.

but are presently under revision. In America, the US EPA has established air quality standards for particulate matter. The primary standards which are set to protect human health are

- 24 hour mean concentration of 150 µg/m³ PM_{10} which is not to be exceeded on more than 0.27% of occasions (1 in 365 days). This was previously set as 260 µg/m³ TSP.

- arithmetric annual mean concentration of 50 µg/m³ PM_{10} which was previously set as 150 µg/m³ annual geometric mean TSP.

Table 6.16 WHO Guideline Values* for Combined Exposure to Sulphur Dioxide and Particulate Matter(a).

	Averaging Time	Sulphur Dioxide µg/m³	Reflectance Assessment: Black Smoke (b) µg/m³	Gravimetric Assessment	
				Total Suspended Particulate (TSP) (c) µg/m³	Thoracic Particles (TP) (d) µg/m³
Short Term	24 hours	125	125	120 (e)	70 (e)
Long Term	1 year	50	50	-	-

Notes
(a) No direct comparisons can be made between values for particulate matter since both health indicators and measurement methods differ. While numerically TSP/TP values are generally greater than those of black smoke, there is no consistent relationship between them, the ratio of one to the other varying widely from time to time and place to place, depending on the nature of the sources.
(b) Nominal µg/m³ units, assessed by reflectance. Application of the black smoke value is recommended only in areas where coal smoke from domestic fires is the dominant component of the particulate matter. It does not necessarily apply where diesel smoke is an important contributor.
(c) TSP: measurement by high volume sampler, without any size selection.
(d) TP: equivalent values as for sampler with ISO-TP characteristics (having 50% cut-off point at 10 µm); estimated from TSP values using site specific TSP/ISO-TP ratios.
(e) Values to be regarded as tentative as they are based on a single study which also involved sulphur dioxide exposure.

* currently under review.

Secondary standards have been established to protect the environment, to increase amenity, etc. These were originally different from the primary standards but have now been set to the same values as for the primary standards.

The Expert Panel on Air Quality Standards has recommended air quality standards for a number of atmospheric pollutants in the United Kingdom, including PM_{10}. The air quality standard recommended by EPAQS (1995) for PM_{10} is 50 µg/m³ as a 24-hour rolling average. Table 6.17 gives the number of exceedences of the EPAQS recommendation as well as those of a number of threshold levels, corresponding to daily mean concentrations of 50 µg/m³, 75 µg/m³, 100 µg/m³ and 150 µg/m³, which have been determined at each of the sites in the national automatic monitoring network for the years 1992 to 1994. Whilst it can be seen that there are a significant number of days at all sites for which the 50 µg/m³ level (both the EPAQS recommendation and the daily concentration) is exceeded, it is clear that there are progressively fewer exceedences of the higher thresholds with only a limited number of exceedences of the 150 µg/m³ threshold. More than one exceedence of the 150 µg/m³ daily level has occurred only at the Belfast site during this period. A further point to note from Table 6.17 is that the number of exceedences of the EPAQS recommended air quality standard for PM_{10} as a 24-hour rolling average is about two times greater than the number of exceedences which would have occured if the standard were expressed as a daily mean concentration of 50 µg/m³. Figure 6.23 shows the mean daily concentration of PM_{10} recorded at all the operational sites plotted for each month for the period from 1992 to 1994. It is evident from this figure that the highest concentrations of PM_{10} and therefore the greatest number of exceedences have tended to occur in the winter months when the dispersion of atmospheric pollution is poorer.

6.7 SPATIAL DISTRIBUTION OF PARTICULATE MATTER

6.7.1 Across an Urban Area

There is an extensive database of black smoke measurements for a range of industrial, residential and commercial site types in urban areas. Although these

Chapter 6 : Concentrations and Trends in Particulate Matter

Table 6.17 Exceedences Recorded between 1992 and 1994 for Various Threshold Values for Daily Average Levels of PM_{10}.

Site	Site First Operational	Year	Number of Days When Daily Mean PM_{10} Levels Exceeded				24 Hour Mean ≥ 50 µg/m³ EPAQS Count	Days	Maximum Daily Mean PM_{10} Level µg/m³
			50 µg/m³	70 µg/m³	100 µg/m³	150 µg/m³			
London (Bloomsbury)	Jan 1992	1992	27	4	0	0	628	43	94
		1993	31	10	0	0	788	57	100
		1994	17	3	0	0	485	39	93
Belfast	Mar 1992	1992	23	12	5	3	552	40	248
		1993	45	13	3	0	1075	80	120
		1994	17	6	3	2	433	32	191
Birmingham (Central)	Mar 1992	1992	21	6	1	0	526	37	131
		1993	25	5	1	0	568	39	102
		1994	15	3	1	0	366	23	113
Newcastle	Mar 1992	1992	21	3	0	0	487	38	72
		1993	14	4	0	0	430	32	79
		1994	23	1	0	0	481	39	77
Cardiff	May 1992	1993	21	3	0	0	492	40	89
		1994	55	13	0	0	1319	91	96
Edinburgh	Oct 1992	1993	2	0	0	0	39	4	66
		1994	1	0	0	0	30	3	62
Bristol	Jan 1993	1993	27	4	0	0	701	52	81
		1994	16	3	0	0	370	30	83
Leeds	Jan 1993	1993	23	4	0	0	594	40	96
		1994	24	7	1	0	549	44	114
Liverpool	Apr 1993	1993	22	6	0	0	449	28	163
		1994	19	5	0	0	471	34	84
Birmingham (East)	Dec 1993	1994	10	2	1	0	234	19	106
Hull	Jan 1994	1994	16	3	0	0	373	30	85
Leicester	Jan 1994	1994	5	0	0	0	178	17	65
Southampton	Jan 1994	1994	7	0	0	0	147	16	63

Note: Entries are only given for those sites which were operational for more than 9 months in a given year.

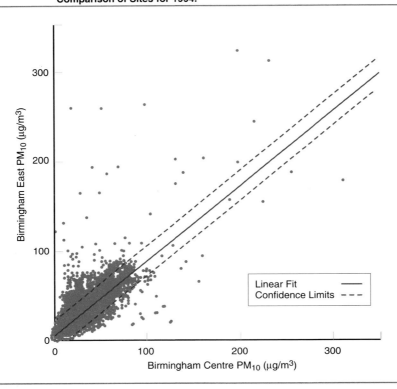

Figure 6.28 Hourly Concentrations of PM$_{10}$ at Two Sites in Birmingham: Comparison of Sites for 1994.

data could be used to infer spatial distributions of particulate matter (as black smoke), the discussion on spatial distributions in this review will be limited to those of PM$_{10}$.

One of the more interesting features of the PM$_{10}$ measurements made at the two sites in Birmingham is the very close correspondence between the concentrations observed. The data given in Table 6.1 for the measurements made in 1994 show the striking similarity not only of the annual mean levels but also in the 98th percentile and peak values. This indicates at least a degree of uniformity in urban concentrations at similar sites. Figures 6.8c and Figure 6.8d show the hourly measurements made at the two Birmingham sites between January 1994 and March 1995 (PM$_{10}$ data for 1995 is provisional). Apart from the peak at the Birmingham East site in May 1994, the PM$_{10}$ concentrations track together and this is further shown in the regression plot in Figure 6.28. A regression analysis indicates that the Birmingham East PM$_{10}$ levels are about 84% of those measured at Birmingham Centre.

The same conclusion was drawn from the PM$_{10}$ measurements made at three sites in the Bristol area. The sites consisted of the national monitoring station in the centre of Bristol, a location in the Avonmouth industrial area about 9 km from the national site and a suburban site at Shirehampton about 1 km from the Avonmouth site. The annual mean concentrations for the 12 month period to March 1995 were respectively 23, 23 and 20 µg/m³ at the three sites. The uniformity across this urban area can also be seen in the daily average results, which are shown for June 1994 and January 1995 in Figures 6.29a and 6.29b. The concentrations follow each other very closely from day to day.

These limited results show little variation in average PM$_{10}$ concentrations across urban areas and are consistent with the uniformity in the annual mean levels observed at the other urban background sites across the UK. A tentative conclusion to draw from these measurements is that one monitoring station can be representative of a particular site type in a given urban area.

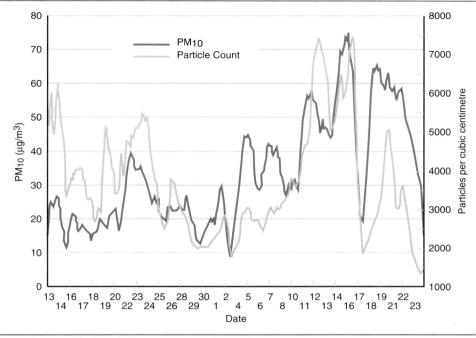

Figure 6.34   24-Hour Rolling Mean PM_{10} and Particle Count from a Site in Birmingham.

Source: Jones and Harrison, 1995.

6.8 ULTRAFINE PARTICLES AND PARTICLE NUMBER COUNTS IN URBAN AIR

The discussion thus far has concentrated solely on the mass of particulate matter measured at the different monitoring sites. There is a view that it is the total number of particles which is of concern rather than the total mass. There have been a number of studies in which the total number of particles as well as the total mass of the particulate matter have been determined.

As indicated in Chapter 2, there are abundant small particles in the atmosphere which contribute substantially to the total particle number count, but represent very little of the total mass concentration because of their tiny volume, and hence mass. These particles are termed ultrafine particles (generally defined as particles less than 50 nm in diameter). A size distribution (Figure 6.31) of particles sampled in the air of Birmingham and expressed in terms of particle number shows that the large majority of particles are smaller than 50 nm, and almost all are below 100 nm diameter. Thus, measures of particle number count, which may be made continuously with a condensation nucleus counter, are a good measure of the ultrafine particle loading.

Especially when close to sources of particle emissions, very large numbers of particles can be counted in urban air. Figure 6.32 shows a time series of particle count measurements taken 3 metres from the busy A38 Bristol Road in Birmingham as part of the DoE Air Quality Research Programme. The count varies rapidly between about 10^3 and $10^6/cm^3$ with an average count of around $180,000/cm^3$. A count made under similar weather conditions at a background site a few hundred metres away revealed a mean particle count of $27,000/cm^3$. Thus particle numbers reduced about six-fold between the roadside site and the background due to dilution with cleaner air and particle coagulation processes which reduce the number count. Comparative measurements of PM_{10} for the same sites indicate that the roadside site has PM_{10} concentrations about 80% greater than the background site. This demonstrates clearly the very large impacts of freshly emitted particles upon the number count at the roadside site. These particles contribute relatively little to PM_{10} mass because of their very small volume. These particle number concentrations may be compared with air over the North Atlantic Ocean, which in the absence of recent pollutant inputs shows typical particle number counts of around 200/cm³.

Measurements of particle count have also been made in central Birmingham at a background site adjacent to the PM_{10} measurements of the Birmingham (Central) Automatic Urban Network site. The majority of particle count data at this site were below 100,000 particles/cm³, although there were occasional excursions above this number, which was the upper limit of the instrument used for particle counting. Figure 6.33 shows moving average PM_{10} and particle count data through one 24 hour period. In general, the two measures of particle abundance are reasonably well related. The variation of particle number count and PM_{10} concentration over a five week sampling period is shown in Figure 6.34. The two exhibit roughly comparable patterns, although the number count starts to rise earlier in the morning than the PM_{10} concentration and does not fall in the afternoon in the way that PM_{10} does at this site. If 24 hour particle count data are correlated with PM_{10} over the same 24 hour intervals, the data show an approximately linear relationship with a correlation coefficient of 0.59.

Measurements made in London by Waller (1976) using electron microscopy revealed concentrations of the order of 10^4/cm³ at background sites, 3-5 x 10^4/cm³ in street samples and up to 16 x 10^4/cm³ in samples taken in tunnels and urban fogs. These measurements suggest that particle number densities have changed little since the 1960s although mass concentrations are expected to have diminished appreciably.

6.9 KEY POINTS

- *There has been a substantial decline in the concentration of particulate matter (as black smoke) over the past 40 years in both urban and rural locations across the United Kingdom. These measurements continue to provide a valuable source of information on the concentrations of particulate matter.*

- *With the introduction of automatic instruments, a sophisticated national urban monitoring network now exists to measure PM_{10}. The measurements have been made mainly in urban background locations and there is now a need to extend the measurement of PM_{10} to other site types.*

- *The annual mean concentration of PM_{10} at the urban background sites in the national Automatic Urban Monitoring Network lie between approximately 20 to 34 µg/m³. Although the annual mean concentrations appear relatively uniform across the United Kingdom, there is a substantial variation from day-to-day, from site-to-site, and from season-to-season.*

- *Analysis of the measurements of PM_{10} with those of other pollutants indicate that strong correlations exist between PM_{10} and CO and also between PM_{10} and NO_x at some locations. Such relationships can be used to infer the concentration of PM_{10} in the absence of direct measurements and for source apportionment.*

- *Attention is increasingly being focussed on the $PM_{2.5}$ size fraction. There are currently very few measurements of $PM_{2.5}$ concentrations in the United Kingdom. Measurements made at Birmingham indicate that the $PM_{2.5}$ size fraction can represent between 30 to 100% of the PM_{10} concentration, the proportion being dependent on the meteorological conditions.*

- *In general, there have been a few or no exceedences of a fixed 24-hour mean PM_{10} concentration of 100 µg/m³ at the sites in the national Automatic Urban Monitoring Network between 1992 and 1994, the exception being the Belfast site where 11 exceedences were observed. On the other hand, there have been widespread exceedences of a daily mean level of 50 µg/m³ at all sites in the network during this period.*

- *The EPAQS recommended standard of 50 µg/m³, 24-hour rolling mean, is exceeded more frequently than the same concentration measured over fixed 24-hour daily periods. Over the period 1992-1994 there were approximately double the number of exceedences of the 50 µg/m³ rolling mean, than of the same concentration expressed as a fixed daily mean.*

- *The number concentration of particles in urban air fluctuates widely, but normally lies between 10^3 and 10^6 per cubic centimetre. This reflects mainly the ultrafine particles which are very numerous but contribute little to PM_{10} mass.*

- *When averaged over 24 hour periods, particle number concentration is correlated with the PM_{10} mass and within a 24 hour period shows a similar diurnal variation.*

- *Particle number concentration is highly elevated close to a busy road, by a factor far greater than the elevation in PM_{10} mass.*

REFERENCES

Anderson HR et al (1995) **The Health Effects of an Air Pollution Episode in London, December 1991**, St George's Hospital Medical School, Department of Public Health Sciences, Cranmer Terrace, London SW17 0RE

Bailey DLR and Clayton P (1980) **The Measurement of Suspended Particulate and Carbon Concentrations in the Atmosphere Using Standard Smoke Shade Methods**, Warren Spring Laboratory Report LR 325 (National Environmental Technology Centre, AEA Technology, Culham Laboratory, Abingdon, Oxon., OX14 3DB).

Ball DJ and Hume R (1977) **The Relative Importance of Vehicular and Domestic Emissions of Dark Smoke in Greater London in the Mid-1970s, the Significance of Smoke Shade Measurements, and an Explanation of the Relationship of Smoke Shade to Gravimetric Measurements of Particulate**, Atmos. Environ., **11**, 1065-1073.

Bower JS, Broughton GFJ, Willis PG and Clark H (1995) **Air Pollution in the UK 1993/94**, NETCEN.

Bower JS, Broughton GFJ, Willis PG, Clark H (1996) **Air Pollution in the UK 1994**, NETCEN.

Clout H (1978) **Changing London**, University Tutorial Press, Yeovil.

Collins G, Jones MR and Harrison RM (1995) Unpublished Data.

Dockery WD, Pope CA, Xu X, Spengler JD, Ware JH, Fay ME, Ferris BG and Speizer FE (1993) **An Association between Air Pollution and Mortality in Six US Cities**, New Engl J Med, 329, 1753-1759.

DoT (1994) **Design Manual for Roads and Bridges**, Department of Transport, Volume 11, Section 3, Part 1 Air Quality.

EC (1980) **Air Quality Limit Values and Guide Values for Sulphur Dioxide and Suspended Particulates**, EC Directive (80/779/EEC).

EPAQS (1995) **Particles,** Department of the Environment Expert Panel on Air Quality Standards, HMSO, London.

Kitto AMN and Harrison RM (1992) **Processes Affecting Concentrations of Aerosol Strong Acidity at Sites in Eastern England**, Atmos. Environ., **26A**, 2389-2399.

International Mining Consultants Limited (1995) **Preliminary Study of Ambient Airborne Particulate Matter**, International Mining Consultants.

Jones MR and Harrison RM (1995) Unpublished Data.

Lee DS and Derwent RG (1995) **The Behaviour of Ammonia in the Atmosphere**, Manuscript in preparation.

Lee DS, Espenhahn S and Baker S (1995) **Base Cation Concentrations in Air and Precipitation in the United Kingdom**, Report AEA/16405532/001/ Final Report prepared for Her Majesty's Inspectorate of Pollution under contract HMIP/CPR2/41/1/146.

Loader A (1994) **Multi-Element Survey: Data Summary 1976/77-1992/93**, Report (AEA/CS/16419032/Z001) prepared for Department of the Environment.

Loader A (1995) **Comparison of PM_{10} and Black Smoke Measurements using Co-located Samplers**, Report (AEA/RAMP/16419046/003) Prepared for Department of the Environment.

LSS (1989) **London Air Pollution Monitoring Network Fourth Report**, Rendel Science and Environment, London.

MAAPE (1992) **Sulphur Dioxide, Acid Aerosols and Particulates**, Second Report prepared by the Department of Health's Advisory Group on Medical Aspects of Air Pollution Episodes, HMSO, London.

MAAPE (1995) **Health Effects of Exposures to Mixtures of Pollutants**, Fourth Report prepared by the Department of Health's Advisory Group on Medical Aspects of Air Pollution Episodes, HMSO, London.

Miller CE and Lewis RH (1996) **Trends in Concentrations of Suspended Particulates in Central Manchester 1980-93**, Environ Technol, in press.

Muir D and Laxen DPH (1995) **Black Smoke as a Surrogate for PM_{10} in Health Studies?** Atmos. Environ., **29**, 959-962.

Pope CA, Thun MJ, Namboodiri MM, Dockery, WD, Evans JS, Speizer FE and Heath CW (1995) **Particulate Air Pollution as a Predictor of Mortality in a Prospective Study of US Adults**, Amer J Respir Crit Care Med, **151**, 669-674.

QUARG (1993a) **Urban Air Quality in the United Kingdom**. First Report of the Quality of Urban Air Review Group, Department of the Environment, London.

Stedman JR (1995) **Estimated High Resolution Maps of the United Kingdom Air Pollution Climate**, Report (AEA/CS/16419035/001) Prepared for Department of the Environment on Contract PECD 7/12/138.

Stevenson KJ, Loader A, Mooney D and Lucas R (1995) **UK Smoke and Sulphur Dioxide Monitoring Networks. Summary Tables for April 1993-March 1994**, Report Prepared for Department of the Environment on Contract PECD 7/12/124.

Waller RE (1967) **Studies on the Nature of Air Pollution. London Conference on Museum Climatology**, International Institute for Renovation of Works of Art; 65-69.

Westminster City Council (1995) Personal Communication.

WHO (1987) **Air Quality Guidelines for Europe. World Health Organisation Regional Office for Europe**, European Series Number 23, World Health Organisation, Copenhagen.

7 The Chemical Composition of Airborne Particles in the UK Urban Atmosphere

7.1 INTRODUCTION

There have been numerous studies of some aspect of atmospheric particle chemistry undertaken on aerosol collected in the UK atmosphere. In this Chapter we seek to review that information and provide an estimate of "typical" UK airborne particle composition. Clearly, composition is variable in both time and space, and since no study has sought to characterise either kind of variation for more than one or two individual components, this point will not be addressed in detail.

The composition of atmospheric particles is influenced by a balance between sources, chemical transformations in the atmosphere, long-range transport effects and removal processes. Particles with a relatively long atmospheric lifetime and no significant localised sources (eg sulphate in fine particles) show quite high spatial uniformity on scales of tens, and possibly hundreds of kilometres. On the other hand, particles with short residence times and/or localised sources (eg quarry dust) show strong spatial concentration gradients. Thus no generalised statements on the degree of uniformity of atmospheric particles are possible. Undoubtedly there are substantial temporal and spatial variabilities in atmospheric particle loadings, and appreciable spatial variations in mean composition. Nonetheless, there are common chemical components which appear at relatively similar concentrations throughout the developed world (Sturges et al, 1989), and within the UK, airborne particle concentrations and composition are not expected to vary greatly from one location to another of the same type (eg urban) elsewhere in the country. There will, however, be local elevations in concentrations of some components, eg marine aerosol at coastal locations.

There are two main sources of information on aerosol composition in the UK atmosphere. Extensive national surveys have been carried out for individual components, chiefly metals and toxic organic compounds. As these studies have only investigated specific trace components of particulate matter, to obtain information on the bulk components use must be made of *ad-hoc* studies carried out by various research groups. This can lead to difficulties in comparing and combining the results obtained by different researchers as different sampling and analytical protocols were used.

A simple comparison of the raw published data from *ad hoc* studies is limited by the following difficulties:

1) There is no overall national strategy behind the selection of sampling sites. The majority of sites for which comprehensive studies have been made are not truly representative of the urban areas in which most people live.

2) There is no common standard for sampler design, so the fraction of particulate matter analyzed in different studies will often contain different particle size groups. It will be shown later that the composition of particles is highly dependent on their size.

3) There are large differences between analytical methods, and in the choice of species analysed.

4) The reported data cover an appreciable time interval and temporal changes in mean concentrations will have occurred over this time.

An additional source of data is from the two national particulate matter monitoring networks: the long-running and extensive National Survey of Air Pollution for black smoke, and the continuous PM_{10} instruments in the Automated Urban Network.

7.11 The Comparability of Data from Size Selective Samples

A frequently quoted distinction between size fractions is that of "fine" particles, aerodynamic diameter < 2.5 µm, and "coarse" particles, 2.5 to 10, or 15 µm. The 15 µm upper limit corresponds to the UK definition for "smoke"; 10 µm to the definition given earlier for PM_{10}. The "fine" fraction can be considered to be "respirable" and includes the "accumulation mode" and is thus the fraction of greatest interest (QUARG, 1993). The "coarse" fraction is largely "inhalable", but clearly includes a substantial fraction which are neither "thoracic" nor "respirable". (See Chapter 3 for definitions). The size fraction in which a species is predominantly found can tell a lot about the source of a species; if it is mostly found in the fine fraction it will probably be anthropogenic in origin. Natural dusts including marine aerosol are in most cases predominantly coarse.

Size selective samplers come in a variety of types with different size ranges employed. The two way division into "coarse" and "fine" is used in the so-called *dichotomous sampler* and has been employed in a study of urban air in Leeds by Clarke et al (1984). Size selection in this sampler is carried out using a virtual impactor, a device unlike the more usual cascade impactor. Because of its direct applicability to health effects studies, the dichotomous sampler is becoming increasingly popular, especially in the US where aerosol sampling has been far more widespread than in the UK. Most of the few UK studies which have taken size-resolved samples have used the cascade impactor which provides samples in several size ranges typically ranging from <0.5 µm to >12 µm. In most other studies of atmospheric particles no size discrimination is employed, making the data obtained less useful for evaluation of health effects. Obviously data from different size ranges are not directly comparable.

When different samplers which nominally capture the same size range are compared they will still not always agree. Studies comparing the High-Volume Sampler with the dichotomous sampler have shown that the Hi-Vol gives higher concentrations of total mass and of some ionic species, especially sulphate (Clarke et al, 1984 and Trijonis, 1983). This may be due, at least in part, to the formation of artefacts on the filter media: gaseous species in the incoming air react with some filter media and with collected particles to produce additional solid matter. Best known is the conversion of sulphur dioxide to sulphate on alkaline glass fibre media.

Even where size selection is used it has been found that the effects of high relative humidity can produce inconsistent results. Particles, especially those with a high sulphate content, will grow when exposed to high relative humidity thus changing their size distribution, a phenomenon investigated in detail by Koutrakis et al (1989). A US study by Keeler et al (1988) investigated the effect this has on dichotomous samplers and found that under extreme conditions 50% of the material usually found in the fine fraction will appear with the coarse. Given that night-time humidities are higher than daytime, and that most public exposure occurs indoors under conditions of low relative humidity, it seems likely that the "fine" fraction will frequently be underestimated by outdoor sampling. During a sampling campaign in Leeds by Clarke et al (1984) it was found that under conditions of high humidity their filters would clog up rapidly so the air was later pre-heated to avoid this problem. Such pre-heating will have altered the size distribution of their samples during the campaign. Pre-heating is also used on the TEOM continuous PM_{10} monitor (Patashnick and Rupprecht, 1991). Conventional methods using Hi-vols and dichotomous samplers do not employ pre-heating, rather the filters are conditioned to standard relative humidity before weighing. This has clear implications for the behaviour of both water content and the collection of volatile components leading to differences in gravimetric concentration. In this case size-distribution will not be affected because the heating is applied after the selective inlet.

All these inconsistencies must be borne in mind when the results of the following sampling campaigns are discussed.

7.2 COMPOSITION OF AIRBORNE PARTICLES

A typical approximate breakdown of UK particulate matter composition may be expressed as follows: ammonium ~5%, sulphate, nitrate and chloride ~30%, carbonaceous material ~40%, metals ~5% and insoluble material (minerals) ~20%. (QUARG, 1993). Figure 7.1 illustrates this graphically. This has been derived by combining data from several studies carried out in different places at different times using different techniques and so cannot be regarded as being especially rigorous. The trace heavy metals and specific organic compounds comprise a tiny percentage of the total mass, and it is these which have been covered by the most extensive sampling campaigns. No survey covering all the components formation. After subtraction of northern hemisphere background CO, the CO to PM_{10} ratio is 15.5 suggesting that less than 20% of PM_{10} in this instance is from primary road traffic emissions.

8.4 INTERPRETATION OF PM_{10} AND $PM_{2.5}$ DATASET FROM NORTH-WEST BIRMINGHAM

A site in North-West Birmingham (Birmingham, Hodge Hill) some 70 metres from the elevated section of the M6 motorway (see Figure 8.7) has been used for simultaneous automated sampling of PM_{10} and $PM_{2.5}$ particles using TEOM instruments. The data collected between October 1994 and March 1995 have been analysed.

The mean concentration at PM_{10} at the site over this period was 16.5 µg/m³, somewhat lower than at the Birmingham Central (22.3 µg/m³) and Birmingham East (21.4 µg/m³) stations of the Automatic Urban Network. The site is clearly traffic influenced as evidenced by the concentration of NO_x, which at 60.9 ppb is higher than either Birmingham Central or Birmingham East, (56.0 ppb and 52.0 ppb respectively). At this site, both PM_{10} and $PM_{2.5}$ correlate strongly with hourly NO_x. The regression equations are as follows:

$PM_{2.5}$ (µg/m³) = 0.114 NO_x (ppb) + 6.54   r = 0.70

PM_{10} (µg/m³) = 0.134 NO_x (ppb) + 11.31   r = 0.70

If the NO_x is taken to be wholly traffic-related, which will be a reasonable assumption at this site as the location is in a heavily trafficked area of a major conurbation, then there is a background of PM_{10} of 11.3 µg/m³ upon which a mean traffic-related concentration of PM_{10} of 5.2 µg/m³ is superimposed. Similarly, for $PM_{2.5}$ the background is 6.5 µg/m³ with a mean concentration of 4.5 µg/m³ of traffic-related particles.

The coarse particle mass can be calculated for each individual hour from the difference between PM_{10} and $PM_{2.5}$. Its mean concentration is 5.5 µg/m³. Coarse particle mass is poorly correlated with NO_x,

PMcoarse (µg/m³) = 0.018 NO_x (ppb) + 5.3
r = 0.33

Based on the above regression equations, road traffic is on average responsible for 41% of $PM_{2.5}$ mass, 32% of PM_{10} mass and 3% of coarse particle mass. This is highly consistent with size distribution measurements of vehicle exhaust which show primary traffic-emitted particles to be very small in size and thus predominantly in the $PM_{2.5}$ fraction. The contribution of road traffic to coarse particles is, as expected, very small.

The correlation of $PM_{2.5}$ with NO_x indicates that on average 4.5 µg/m³ arises from road traffic and 6.5 µg/m³ comes from other sources. The mean annual average concentration of secondary sulphate and nitrates predicted for Birmingham from the map in Figure 8.2 is 10 µg/m³. Approximately 80% of sulphate and nitrate mass is expected to be in the $PM_{2.5}$ fraction, and applying a seasonal adjustment (Figure 8.3), this translates to a secondary $PM_{2.5}$ contribution over the months October to March of 7.0 µg/m³, almost identical to the non-traffic related $PM_{2.5}$ of 6.5 µg/m³. It may thus be concluded that $PM_{2.5}$ is accounted for almost wholly by traffic exhaust-emitted and secondary particles. On the other hand, coarse particles, which account for approximately one third of the PM_{10} mass in this case, are not associated with vehicle exhaust emissions, although they may arise in part from wear of tyres and road surfaces and resuspension of dust by passing traffic.

8.5 IMPLICATIONS FOR THE CONTROL OF ATMOSPHERIC CONCENTRATIONS OF PM_{10}

Figure 8.8 shows hourly concentrations of coarse particles as a function of PM_{10} concentrations. If we define elevated concentrations of PM_{10} as above 50 µg/m³, then there are five hours in which coarse particles make the major contribution to raised levels of PM_{10}. For the other 150 hours of elevated PM_{10}, whilst the PM_{10} concentration increases to 170 µg/m³, the coarse particle mass is almost level at around 20-25 µg/m³. Thus by far the majority of episodes (97% in this dataset) are those associated with excursions in

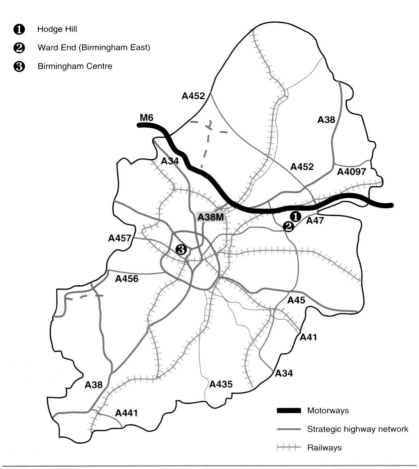

Figure 8.7 Locations of Air Quality Sampling Sites in Birmingham.

Source: Adapted from Birmingham City Council - The Environment in Birmingham: 1993.

$PM_{2.5}$ concentrations. *Thus, if elevated concentrations of PM_{10} are to be avoided in the Winter (when most episodes occur), the primary requirement will be control of the sources of $PM_{2.5}$.*

As noted above, the main sources of $PM_{2.5}$ are (a) road transport emissions and (b) secondary aerosol. A clear indication of the relative importance of these two sources at the Birmingham, Hodge Hill site in the winter of 1994-95 may be gained from Figures 8.9 and 8.10 which relate PM_{10} and $PM_{2.5}$ to NO_x, itself a good indicator of local vehicle emissions. There is a group of some seven points where at concentrations of NO_x below 250 ppb, the concentration of particulate matter exceeds 150 µg/m³. These are episodes where substantially elevated $PM_{2.5}$ is not associated with NO_x and these are interpreted as due to a local source, probably industrial in character. For the remaining data, $PM_{2.5}$ generally relates closely to NO_x. There are a few points where $PM_{2.5}$ exceeds 50 µg/m³ which lie well above the regression line and these are interpreted as having a major contribution from secondary particles. The vast majority of points, however, lie very close to the regression line indicating that road traffic is far the major cause of elevated $PM_{2.5}$, and by implication of PM_{10}, at this site. It should be noted that this is not an abnormally polluted site since it has PM_{10} concentrations lower than Birmingham Central and Birmingham East, and a mean NO_x concentration only slightly higher.

The very clear implication of this dataset is that in the winter period the control of elevated concentrations of PM_{10} can only be achieved by reduction of road traffic emissions. Thus, whilst control of SO_2 and NO_x emissions will have a benefit upon concentrations of

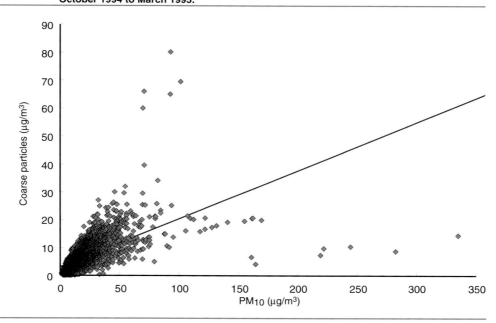

Figure 8.8 Plot of Hourly Coarse Particles versus PM$_{10}$ for Birmingham, Hodge Hill; October 1994 to March 1995.

secondary PM$_{2.5}$ and PM$_{10}$ as indicated in Chapter 5, this will not have any substantial influence upon winter episodes of PM$_{10}$ pollution.

8.6 SUMMER DATA

Data from the months April to July 1995 have been analysed by the same methods as for the winter data.

In the summer period, the mean PM$_{10}$ at the Hodge Hill site was 24.7 µg/m³, the same as at Birmingham Centre (24.7 µg/m³) and higher than at Birmingham East (18.3 µg/m³). However, the maximum hourly PM$_{10}$ was highest at Birmingham Centre (193 µg/m³), followed by Birmingham East (137 µg/m³) and Hodge Hill (118 µg/m³). Mean NO$_x$ was highest at Hodge Hill (49.1 ppb) compared with Birmingham Central (31.5 ppb) and Birmingham East (22.5 ppb).

The mean concentration of PM$_{2.5}$ at Hodge Hill was 13.1 µg/m³, representing only 53% of PM$_{10}$. The two were quite strongly correlated:

$$PM_{2.5} = 0.51\ PM_{10} + 0.57\ (\mu g/m^3) \qquad r = 0.85$$

Correlations of PM$_{10}$ and PM$_{2.5}$ with NO$_x$ were positive, but weak in comparison with the winter data.

Figures 8.11 and 8.12 show the relationship between fine (PM$_{2.5}$) and coarse (PM$_{10}$ minus PM$_{2.5}$) particles respectively and PM$_{10}$. The data show broad scatter across an envelope with outer bounds of around 85% coarse particles at one extreme and 85% fine at the other extreme. The high abundance of coarse particles, which exceeds that found in other studies in Birmingham, is probably associated with the unusually dry weather over much of this period. Concentrations of hourly PM$_{10}$ above 50 µg/m³ are associated both with coarse and fine particles. Exceedences of 90 µg/m³ are in all but one instance associated with a predominance of PM$_{2.5}$ and the implications for control of excursions are the same as for the winter period, with the caveat that occasionally episodes of coarse dust may be a problem.

8.7 MEETING THE EPAQS RECOMMENDED STANDARD

Table 6.17 lists exceedences of various thresholds in 24 hour average concentration of PM$_{10}$. There are currently many exceedences of 50 µg/m³ over 24 hours, but far fewer exceedences of 100 µg/m³ and virtually none of 150 µg/m³. Exceedences of the 50 µg/m³ threshold are more frequent for the rolling 24-hour mean than the fixed (midnight to midnight)

Figure 8.9 Plot of Hourly Fine Particles (PM$_{2.5}$) versus NO$_x$ for Birmingham, Hodge Hill; October 1994 to March 1995.

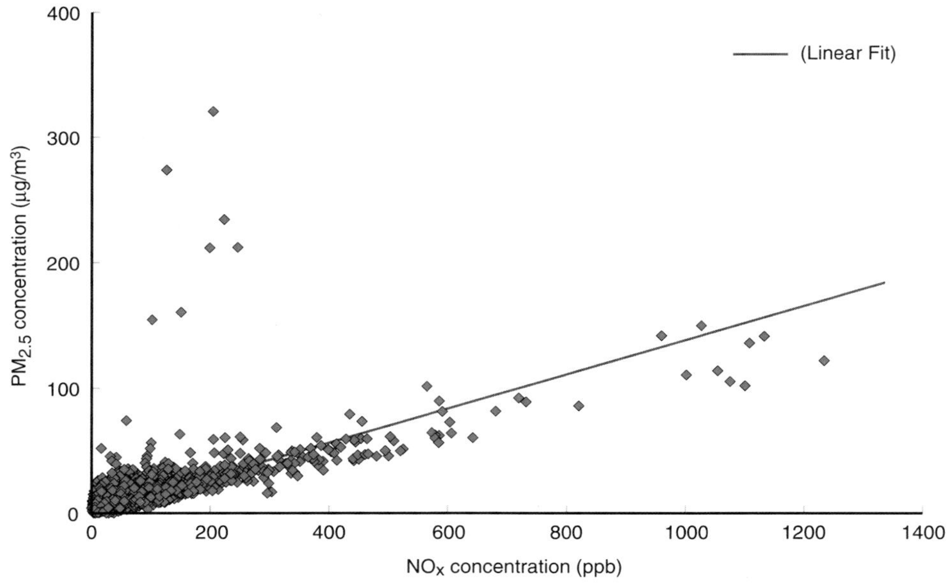

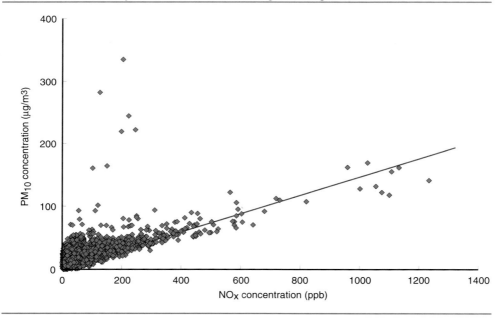

Figure 8.10 Plot of Hourly PM$_{10}$ versus NO$_x$ for Birmingham, Hodge Hill; October 1994 to March 1995.

means. However, it is clear that a reduction of all 24-hour concentrations currently in the 50-100 µg/m³ band to below 50 µg/m³ would reduce dramatically the number of exceedences of the EPAQS recommended limit.

For the winter data, it is possible to estimate approximately the extent of control of the major source of PM$_{10}$ during episodes of high concentration, (ie vehicle emissions), in order to reduce a concentration of 100µg/m³ to 50 µg/m³. The Birmingham (Hodge Hill) data indicate that during episodes of high PM$_{10}$, the coarse particle fraction

9 Effects of Airborne Particulate Matter

9.1 HEALTH EFFECTS OF NON-BIOLOGICAL PARTICLES

A detailed review of the possible effects of particulate air pollution on health has recently been completed by the Department of Health's Committee on the Medical Effects of Air Pollutants (COMEAP, 1995). These findings are summarised here. Readers are referred to the COMEAP report for details of the evidence considered and the detailed justification for the conclusions presented.

9.1.1 Summary of COMEAP Report

The evidence considered in the COMEAP report is concerned with effects on health of "suspended particulate matter" (SPM), comprising solid (soluble or insoluble) or liquid material present in the air in particles small enough to remain in suspension for some hours or days and liable to travel considerable distances from the source. This corresponds approximately with the size range of particles, typically less than 10 µm diameter, capable of entering the respiratory tract and reaching the deeper parts of the lungs.

SPM is not a defined entity: both chemical and physical characteristics vary widely with respect to source, location and time. There is no clear evidence as to whether effects on health are related to certain components or whether they represent non-specific effects of inhaled particles. Other pollutants or other factors in the environment, for example, changes in temperature, can affect the same health endpoints. It has proved difficult to do representative studies of the effects of the ambient aerosol under controlled conditions owing to the impossibility of recreating the precise particle mixture found in outdoor air. Studies of the effects of well defined small particles could be done and these would contribute useful baseline data; unfortunately few such studies have been reported. Because of this, there is little evidence other than from epidemiological studies. Some insight into possible mechanisms of effect have come from animal studies though no completely satisfactory explanation for the findings of epidemiological studies has yet been produced.

PM_{10} represents the size range of particles likely to pass the nose and mouth, $PM_{2.5}$ represents more closely the size range of particles able to reach the deeper parts of the respiratory tract. A significant proportion of the mass of material collected as PM_{10} is less than 2.5 µm in diameter. Particles of about 0.5 µm diameter are least likely to be deposited in the respiratory tract. Exercise may have a variable effect on particle deposition depending on particle size and pattern of respiration. The estimated typical adult deposition of particles in the PM_{10} range is about 250 µg/day: in terms of traditional toxicology of inhaled particles a small dose.

Some experimental studies, mainly in animals and at high concentrations, show that mixtures of particles with pollutant gases such as SO_2, NO_2 or O_3 have effects greater than those of the components separately, though it is not clear whether these results would apply at the low concentrations normally present in the environment. The adsorption of such gases, or of metals or acid-sulphates, on the surface of particles may be important. Animal studies also indicate a potentially important role for ultrafine particles (< 0.05 µm diameter). They are cleared only very slowly from the lung and they can penetrate the pulmonary interstitium inducing inflammatory responses. It has been suggested that allergenic material may be carried into the lung in association with particles though evidence is lacking.

Extensive studies of the effects of well characterised particles (eg, acid sulphates and sulphuric acid) have been carried out in volunteers. It has been demonstrated that at the concentrations likely to be experienced in the UK no effects would be expected in normal individuals. Exposure to concentrations of acid particles and gaseous pollutants has been shown to lead to an enhanced effect though the findings are not consistent. Subjects with asthma show enhanced non-specific reactivity following sulphuric acid challenge, but not, apparently, following sulphate challenge. However, all these changes in either lung volumes or bronchial responsiveness are, in general, small. The reasons for this may be in part methodological, but the evidence suggests that there is unlikely to be a large effect of acid or sulphate challenge on human subjects.

On the basis of findings from studies of pollution episodes in the 1950s to 1960s, when concentrations of suspended particulate matter, together with SO_2, were very high, it was not anticipated that adverse effects on health would be demonstrable at the much reduced concentrations of more recent years. However, a series of intensive studies, mainly in the United States, has demonstrated small changes in a wide range of health indicators, showing clearer associations with concentrations of particles than with other pollutants.

To some extent, the new findings have been dependent on the introduction of advanced statistical computing techniques, handling large data sets covering many variables. Such methods have passed into general use during the past 10-15 years. These procedures have been examined carefully and they are considered to be appropriate and the results are robust to a variety of statistical methods. It is unlikely that the findings of recent epidemiological studies are an artefact of the statistical methods employed.

Indicators of ill-health examined in relation to acute effects have included day-to-day variations in mortality, hospital admissions, emergency room visits, time off school or work, respiratory symptoms, exacerbations of asthma and changes in lung function. Of these various indices, the measurement of mortality is the most certain. In addition, effects on mortality have been particularly well studied and the results have been generally more consistent than those regarding the other indices. In most, associations with SPM, expressed at least in the more recent studies as PM_{10}, have been observed, within a relatively low range of concentrations of similar magnitude to those found in the UK today. It does not necessarily follow that the same types or sizes or components of particles are involved in these diverse effects. The composition of particles monitored as PM_{10} can vary widely from area to area and with time. However, the size of the estimated effects, particularly as regards effects on mortality, does not vary greatly with location.

Short-term variations in these health indicators have been demonstrated to be influenced by factors such as temperature and (even minor) epidemics of infections such as influenza, as well as possible effects of other pollutants, including SO_2 and ozone. Considerable efforts to control statistically for the effects of confounding factors have been made in recent epidemiological studies. For example, in the most recent studies in the United States, possible effects of air pollution have only been examined once all other identifiable variations in the health indicators have been accounted for. The likely success of the methods used inevitably depends on how well the confounding factors have been characterised and on this point there remains room for debate. Overall, control for confounding factors, in the more recent studies, is accepted as good. The possibility of some unmeasured confounding factor playing a role cannot be dismissed, but this is not considered likely.

Though the results of most epidemiological studies are described as providing exposure-response relationships, it should be remembered that, in general, they provide information on ambient concentration-response relationships across the population studied. In any population the range of personal exposure and consequently dose is likely to be considerable and monitoring the ambient concentration of a pollutant will provide only an estimate of exposure.

There is no clear indication that effects on health are restricted to specific types of particles. Findings from epidemiological studies have demonstrated associations with particles in a wide range of circumstances, including those in which primary emissions from motor vehicles, industrial sources or coal fires or where secondary aerosols derived from gaseous emissions (SO_2 and NO_x from power stations and other stationary sources with further NO_x from traffic) are important. It is unlikely, however, that coarse, wind-blown particles have a significant effect upon health.

It is well established from the reported studies that those with pre-existing respiratory and/or cardiac disorders are at most risk of acute effects from exposure to particles. It has been suggested that acute effects occur when air pollution worsens an acute condition such as a respiratory infection or heart attack in people with pre-existing chronic disease. There is no convincing evidence that healthy individuals are likely to be significantly affected by levels of particles found in ambient air in the UK.

In most of the statistical analyses of epidemiological studies, relationships have been expressed as regression equations from which the magnitude of effects linked with changes in pollution can be judged. An example, from a study in Detroit, indicates that a 30 µg/m³ increase in PM_{10}, as a 24 hour average, was associated with a 3.5% increase in hospital admissions for pneumonia. This, of course, does not imply that exposure to such pollution initiated the illness: it should be considered as causal only through predisposing to, or worsening, the illness and precipitating the need for admission to hospital.

No evidence for thresholds of effect has appeared from the studies so far reported, though it should be noted that several studies are not powerful enough to examine the issue critically at low concentrations.

The principal question to consider in reviewing the rapidly expanding literature on effects of SPM is whether the statistical associations demonstrated indicate a causal role. There is certainly a remarkable degree of consistency and coherence in the direction and magnitude of findings from a diversity of studies, carried out in different localities in the United States and elsewhere, with a range of different health indicators and varying sources of pollution. COMEAP considers the reported associations between levels of particles and effects on health principally reflect an actual relationship and not some artefact of technique or the effect of some confounding factor. The indications that the association is likely to be causal are certainly strong.

The only major difficulty in reaching any firmer conclusion about causality is the lack of established mechanisms of action. The mass of SPM associated with adverse effects is very small, and while there is evidence relating to acute effects of some components, the fact that in epidemiological studies similar effects have been reported in localities with different types of suspended particulate matter suggests that there may be a non-specific effect of particles. The effects have not been explained in terms of the results of conventional inhalation toxicology studies but carriage of material on particle surfaces could play a role. It has been suggested, but by no means proven, that ultrafine particles (< 0.05 µm) may play a role. Such particles would be more prominent close to sources, before they had time to coalesce into the more stable accumulation mode, and they would represent only a small proportion of the mass of material measured as PM_{10}, though they would represent a high proportion of the number of particles present. Although reported studies indicate a range of effects, from small changes in ventilatory function or exacerbations of asthma through to increases in deaths among the elderly or chronic sick, it does not necessarily follow that the same components would be involved in each effect.

9.1.1.1 *Implications for Public Health in the UK*

COMEAP concluded that, in terms of protecting public health, it would be imprudent not to regard the associations as causal and also that the findings of the epidemiological studies of the acute effects of particles, which have been conducted in the US and elsewhere, can be transferred to the UK at least in a qualitative sense. However, it was considered that there are insufficient UK data available to allow direct extrapolation and reliable estimation of effects in the UK.

It would be possible, but unwise, to take a weighted average of the results of the published studies and apply this to conditions in the UK. The coefficient produced by Schwartz (1994) with regard to effects of particles on mortality was 1.06 (CI: 1.05-1.07) for a 100 µg/m³ change in TSP. No similar figure based on structured meta-analysis and providing confidence intervals is available with respect to PM_{10}. Because of this, COMEAP felt it would be unwise to offer a single coefficient with regard to effects on mortality or other indices of ill health. The reader is referred to the COMEAP report where a collection of estimates derived from a wide range of studies is presented. It should be recalled that estimates based on studies reported in this collection are likely to provide only a first approximation to the actual effect in the UK. COMEAP recommend that studies should be undertaken urgently to allow better quantitative predictions to be made. In comparison with other uncontrolled factors, eg, variations in temperature, the effects of day-to-day variations in levels of particles are small.

9.1.1.2 Chronic Effects

With regard to indicators of possible chronic effects on health as a result of exposure to particles, in the UK, there has been a long-established association between the prevalence of, and mortality from, chronic respiratory disease (bronchitis) and exposure to the SO_2/particle complex associated with coal-burning. The role of the particulate component has remained uncertain and other factors, notably smoking, have been shown to be of major importance. Disentangling the roles of multiple factors operating throughout life has proved to be a difficult task, but in some recent US studies, it has been possible to follow selected population groups for a number of years, relating death-rates among them to local pollution levels. Associations have been found to be closer with concentrations of particles than with those of other pollutants, but in view of uncertainties about the possible relevance of (higher) exposures to pollution earlier in life and the difficulties in fully adjusting for confounding factors, quantitative assessments are in doubt. The lower prominence given to this aspect of the effects of particles on health in the COMEAP report reflects the lack of data rather than the potential importance of possible effects on the public health.

Evidence regarding the effects of long term exposure to particles on health is less well developed than that regarding the acute effects. The possibility of confounding in such studies is considerable and it is difficult to estimate the exposures of individuals over relevant time periods. Here again the results of recent US studies are probably transferable to the UK in a qualitative sense though confidence in the accuracy of the predictions is lower than with regard to the acute effects of particles. Though the evidence is limited COMEAP advised that it would be prudent to consider these associations as causal.

There is little evidence to show that exposure to atmospheric particles contributes significantly to the burden of cancer in the UK. The presence of genotoxic carcinogens in particles means that such a contribution cannot be ruled out, though it is likely to be exceedingly small.

9.1.1.3 Conclusions

The COMEAP report ended with the following brief conclusions:

"The Committee considers the reported associations between levels of particles and effects on health principally reflect an actual relationship and not some artefact of technique or the effect of some confounding factor.

In terms of protecting public health it would be imprudent not to regard the demonstrated associations between levels of particles and effects on health as causal.

We find it difficult to reach a firmer conclusion about causality due to the lack of any established mechanism of action.

We believe that the findings of the epidemiological studies of the acute effects of particles, which have been conducted in the US and elsewhere, can be transferred to the UK, at least in a qualitative sense.

It as accepted that insufficient UK data are available to establish the reliability of quantitative predictions of the effects of particles upon health in the UK.

We consider that results of recent US studies of the effects of long-term exposure to particles are probably transferable to the UK, though confidence in the accuracy of the predictions is lower than with regard to the acute effects. Although the evidence is limited, we advise that it would be prudent to consider these associations as causal.

There is no convincing evidence that healthy individuals are likely to be significantly affected by levels of particles found in ambient air in the UK."

Colbeck I and Harrison RM (1984) **Ozone-Secondary Aerosol-Visibility Relationships in North West England**, Sci Tot Environ, **34**, 87-100.

Committee on the Medical Effects of Air Pollutants (1995) **Health Effects of Non-Biological Particles**, Department of Health, HMSO, London.

Gomez B and Smith CG (1987) **Atmospheric Pollution and Fog Frequency in Oxford, 1926-1980**, Weather, **42**, 98-106.

Horvath H (1995) **Estimation of the Average Visibility in Central Europe**, Atmos Environ, **29**, 241-246.

IPCC (1996) **IPCC Second Assessment Report**, in Press.

Lee DO (1990) **The Influence of Wind Direction, Circulation Type, and Air Pollution Emissions on Summer Visibility Trends in Southern England**, Atmos Environ, **24**, 195-201.

Lee DO (1994) **Regional Variations in Long-term Visibility Trends in the UK, 1962-1990**, Geography, **79**, 108-121.

QUARG (1993) **Urban Air Quality in the United Kingdom**, QUARG, London.

Schwartz J (1994) **Air Pollution and Daily Mortality: a Review and Meta-Analysis**, Environ Res, **64**, 36-52.

Williams ID and McCrae IS (1995) **Road Traffic Nuisance in Residential and Commercial Areas**, Sci Tot Environ, **169**, 75-82.

10 Conclusions

The study of airborne particulate matter has lagged behind research on most of the other common air pollutants. This has probably happened because particulate matter was not perceived as a significant threat to health once the extreme pollution characterised by the London smogs was abolished. The other problems associated with airborne particulate matter, such as soiling and visibility degradation were not perceived as major issues in comparison to health concerns relating to pollutants such as sulphur dioxide, nitrogen dioxide and ozone. However, the recent epidemiological work pioneered by the Harvard School of Public Health, but now being repeated around the world, has built a strong case for believing that fine airborne particles have a serious detrimental influence on health. This has led to a burgeoning of interest in particulate pollutants, although the base of data has been severely limited.

For reasons which were sound at the time, airborne particulate matter in the UK atmosphere has traditionally been measured by the Black Smoke method. This nowadays gives no useful indication of the mass of particles per unit volume of air, and it was a far-sighted decision by the Department of Environment to introduce automatic gravimetric TEOM monitors for particulate matter when the Enhanced Urban Network (now part of the Automatic Urban Network) was established, starting in 1991. This Network, which is still growing, has provided invaluable high quality data upon the concentrations of PM_{10} particles in our major towns and cities. The data available to the policy makers and research workers from this network, to be made available in due course to the general public, probably equals that of any other country in the world and far exceeds the quality of data available in many other developed countries in terms of the time response and speed of availability to the user.

Airborne particles in the UK atmosphere arise from a wide variety of sources. Receptor modelling methods, which depend upon hour-to-hour and day-to-day variations in the concentrations and chemical composition of the airborne particles, give a useful indication of the diverse range of sources. These show road transport to be a very major source of particles in urban air, both through exhaust emissions and coarse dust resuspended from the road surface by atmospheric turbulence induced by the passage of traffic. Traffic also generates particles from the wear of rubber tyres and of the road surface itself. Both sea spray and de-icing salt lead to the presence of sodium chloride particles in the atmosphere, even at locations remote from the sea. The other main source of particles arises from the emission of sulphur dioxide, mostly from power stations and other industry, and of nitrogen oxides from road traffic, power stations and other combustion sources. These gases are oxidised in the atmosphere to form sulphuric and nitric acids which are neutralised with atmospheric ammonia to form the so-called secondary aerosol of mainly ammonium sulphate and ammonium nitrate. Because of its rather slow formation and long atmospheric lifetime, concentrations of secondary particles are far more uniform across the UK than the primary emissions from road traffic, which are concentrated mainly in urban areas. The European continent contributes significantly to the airborne loading of secondary particles in the UK atmosphere. Concentrations of secondary aerosol are highest in the South-East of the country, diminishing towards the North and West. Road transport contributes to secondary aerosol concentrations through emissions of NO_x. In London, a very high proportion of PM_{10} is accounted for by the sum of vehicle exhaust emissions and secondary particles.

Airborne particles vary greatly in their size and this influences their atmospheric behaviour. Particles emitted from combustion sources and formed from gas to particle conversion are initially very small in size (around 0.02 μm diameter) and very numerous. These particles have the greatest influence on the atmospheric particle number count, which in polluted locations often exceeds 100,000 particles per cubic centimetre. These very fine nucleation mode particles rapidly coagulate with other particles in the air to form the accumulation mode, comprising particles of intermediate size which are capable of existing in the atmosphere for 10 days or so, which can therefore be transported over very large distances by the wind. Most particles arising from human activity fall into these two size categories which have the greatest implications for human health. Airborne dusts produced by mechanical disintegration processes such as grinding and milling, break-up of soil surfaces by the wind and formation of sea spray from breaking waves tend to be appreciably larger in size and form the coarse particle fraction within PM_{10}. Whilst such particles arise in part from human activity, such as dusts resuspended from the road surface by passing traffic, a major proportion is probably natural in origin and hence very difficult to subject to controls. Indeed, it is only because the UK has a relatively wet climate and much of the ground is covered in vegetation that airborne concentrations of wind-blown dust are not much higher.

Emissions inventories are well able to quantify sources such as road traffic exhaust and power station emissions. They also include such activities as quarrying and the generation of particles from tyre wear; the associated emission factors must be far less certain, however. The inventories attribute an appreciable proportion of airborne particulate matter nationally to road traffic emissions and show that in urban areas the influence of road traffic is very much greater. Emissions from diesels make up far the major part of the total particle mass from road traffic; the modern catalyst-equipped petrol engine is currently the cleaner option with respect to PM_{10} emissions. The inventories are not well able to accommodate secondary particles formed in the atmosphere, but high quality numerical models of atmospheric chemistry and transport are available, which not only give a good capability for predicting current concentrations of secondary components in the atmosphere, but can also predict the impact of future controls on SO_2 and NO_x emissions upon airborne concentrations of secondary PM_{10}.

Control strategies for airborne particulate matter must address the sources of particles contributing to episodes of elevated concentrations. In some areas, these include coal burning, and locally may include construction and demolition activities as well as industrial operations like iron and steel production and minerals extractions. In the great majority of urban areas, however, vehicle emissions, predominantly of fine particles ($PM_{2.5}$) are the dominant source. Simultaneous measurements of PM_{10} and $PM_{2.5}$ show that in winter road traffic contributes little to coarse particle concentrations in the atmosphere, and also that coarse particles rarely contribute significantly to raised concentrations of PM_{10} in the urban atmosphere. Coarse particles are also of limited atmospheric lifetime, and hence travel distance, and may thus give rise to local problems, but have little impact on a city-wide or national basis. The sources are largely natural and/or diffuse, and control strategies need careful consideration in the light of this.

There are a number of reasons why control strategies for PM_{10} in urban areas should concentrate on the fine $PM_{2.5}$ fraction which include:

- The most severe excursions in PM_{10} occur in the winter months when $PM_{2.5}$ is the predominant fraction.

- As noted above, it is excursions in $PM_{2.5}$ which lead to the majority of elevated concentrations of PM_{10}, especially in the winter months.

- The sources of $PM_{2.5}$ are well defined and are controllable.

The work in which $PM_{2.5}$ and PM_{10} have been measured simultaneously over a period of many months clearly indicates that road traffic exhaust emissions and secondary aerosol together can account for essentially all of the atmospheric loading of $PM_{2.5}$. It follows that these are the sources which should be the main target if controls on PM_{10} are required.

Using data upon winter concentrations of PM_{10} it is possible to calculate the extent of control required to reduce 24-hour PM_{10} concentrations of 100 µg/m³ to meet the EPAQS recommended standard of 50 µg/m³. The reasoning has been set out in Chapter 8 and leads to an estimated requirement of about 60% control of vehicle-emitted particulate matter if the majority of central urban background sites are not to exceed 50 µg/m³, or are to exceed it only very infrequently. The projected reduction in emissions of particulate matter from road traffic by the year 2010 are about 50% but this gives no ground for complacency. Firstly, the projected reductions in particulate matter emissions assume that renewal of the vehicle fleet will continue at a rate comparable to that in the past, and that new vehicles will conform to the EC emissions limits and continue to do so even as they age. Secondly, the dataset from which current airborne concentrations are taken comes from measurements at central urban background locations. To meet the EPAQS standard at roadside sites, where concentrations are typically double the urban background, and during extreme episodes will require a far greater degree of control. For example, to have limited 24-hour average PM_{10} to below 50 µg/m³ in the December 1991 pollution episode in London is estimated to require a reduction of over 80% in road traffic exhaust emissions.

Data based on fixed 24-hour measurements of PM_{10} listed in Table 6.17 indicate that measures which effect a reduction of all 24-hour PM_{10} concentrations currently in the 50-100 µg/m³ band to below 50 µg/m³ would have a dramatic influence upon air quality. Exceedences of 50 µg/m³ over fixed 24-hour periods would drop from 442 to just 5 in the 1992-1994 data from the network listed in Table 6.17 if Belfast, which is exceptional due to the continued combustion of solid fuels in domestic premises, is omitted. This requires a 50% cut in peak concentrations of PM_{10}, although to bring all rolling 24-hour concentrations below the 50 µg/m³ recommended by EPAQS would require a larger reduction. Whilst the majority of exceedences of 100 µg/m³ (24 hour average) occur in winter, concentrations in the 50-100 µg/m³ band occur in all months of the year. In summer, the seasonal average contribution of vehicle emissions to PM_{10} is considerably less than that in winter, and the significant correlation of peak hourly PM_{10} with peak hourly ozone clearly indicates the importance of photochemically produced secondary particles at this time of year. Coarse particles can also make a substantial contribution in summer. Numerical projections using the HARM model indicate that a 39% reduction in secondary particles is likely to be achieved by the year 2010 on the basis of agreements already in place for the control of sulphur and nitrogen oxides. The background PM_{10} due to coarse particles is typically greater in summer than in winter, and although our understanding of the episodicity of PM_{10} pollution due to secondary particles is not as complete as for vehicle-emitted particles in winter, it seems certain that a level of control of secondary particles well in excess of 39% will be needed to reduce summer PM_{10} to below 50 µg/m³ given the high background of coarse particles. A complementary strategy would involve control of coarse particles also, but current understanding of the sources and hence implications for control is insufficient in relation to coarse particles.

Detailed analyses of the air quality data clearly show the immense importance of road traffic emissions in influencing $PM_{2.5}$ and PM_{10} concentrations when these are elevated. In winter, a dramatic reduction of hourly exceedences of 50 µg/m³ of PM_{10} could be achieved solely by limiting road traffic exhaust emissions. Currently, diesel vehicles, in particular trucks and buses, are the main contributor to PM_{10} emissions from this sector. In our Second Report, the results of predictive calculations of future emissions of particulate matter from road transport were presented. These indicated the importance of control of diesel vehicles in reducing emissions of particulate matter from road traffic. The revised projections of future emissions included in this report show that measures currently in place will not deliver a sufficient reduction in primary emissions to ensure that all concentrations fall within the EPAQS recommended limit, even by the year 2010. These projections are based upon a modest level of market penetration by diesel cars, and any increase in market share for diesels would inevitably make matters worse as the current technology diesel car emits far more particulate matter than the modern petrol car, and although technological development of the diesel will narrow the gap, there is no current prospect of the diesel improving beyond the petrol car. Additionally, the diesel car emits more nitrogen oxides than the near-equivalent petrol vehicle with a three-way catalyst and hence the impact of more diesel cars in the parc upon future secondary particle concentrations will also be deleterious.

In its eighteenth report on transport and the environment, the Royal Commission on Environmental Pollution recommended the implementation of the most stringent controls practicable on particulate matter in the Stage III emission limits to be introduced by the European Commission in the year 2000. The Royal Commission also drew attention to the problems of simultaneous control of both particulate matter and nitrogen oxides from the diesel and recommended that limits on one pollutant should not be at the expense of the other. It is clear from the data and analyses presented in this report that stringent additional controls both on particulate matter from road transport, and upon the emissions across Europe of sulphur and nitrogen oxides responsible for the formation of secondary particles in the atmosphere, will be essential if UK urban particulate matter concentrations are to be reduced in line with the recommendation of the Expert Panel on Air Quality Standards. Additionally, in some localities, further controls on burning of coal and other solid fuels, and of emissions from major point sources will also be required.

Research Recommendations

Inventories

1. Further studies of emission factors for particulate matter from road vehicles, especially heavy duty vehicles are needed. This should include tyre and brake wear as well as exhaust emissions. Where possible, on-the-road measurements are preferable.

2. Size differentiated emission factors for other sources of primary particulate matter such as construction work, quarrying, agriculture and resuspension from road surfaces should be determined under typical UK conditions.

3. Inventories of primary PM_{10} are needed for a greater number of urban areas.

Monitoring and Chemistry

4. A number of PM_{10} monitoring stations should be located in rural areas to establish rural concentrations and to evaluate particulate matter concentrations in air arriving from the European continent.

5. Monitoring of $PM_{2.5}$ should be carried out at a number of Automatic Urban Network stations.

6. Measurements of particle number count should be carried out on an experimental basis at sites including urban, rural and roadside locations.

7. Measurements of size distribution of particles in the UK atmosphere in terms of particle number and mass, and for individual chemical constituents, should be made.

8. Studies of personal exposure to airborne particles should be conducted, including co-exposure to particles and airborne allergens.

9. Co-located measurements of TSP, PM_{10}, $PM_{2.5}$ and black smoke should be conducted at selected sites to establish inter-relationships in relation to particle chemical composition.

10. Studies of the surface chemical composition of particles are relevant to a full comprehension of their health impact, and are hence recommended.

11. Studies of the spatial distribution of PM_{10} and $PM_{2.5}$ across urban areas are needed.

Receptor Modelling and Source Apportionment

12. Detailed studies of the full chemical composition of representative samples of PM_{10} are required.

13. Daily samples of PM_{10} and $PM_{2.5}$ should be analysed for selected chemical species indicative of specific sources over at least a year at representative sites.

14. Studies should be undertaken of the composition and sources of coarse particles, whose origins are currently poorly defined.

15. Studies of the influence of meteorological factors upon concentrations of PM_{10} and its specific components are recommended.

Effects of Particulate Matter

16. The role of urban particles in influencing cloud and fog formation and visibility deserve attention.

17. Further detailed studies of the health effects and mechanism of action of PM_{10} in humans are required. Specific recommendations appear in the report of COMEAP.

Glossary of Terms and Abbreviations

Accumulation mode
A part of the size spectrum of airborne particles, between approximately 0.1-2 µm diameter, in which particles have a long atmospheric lifetime

Aeroallergens
Airborne antigens that cause allergy in a sensitive individual, e.g. pollen, house dust mite

Aerodynamic diameter
The aerodynamic diameter of a particle is the diameter which the particle would have if it were to be spherical in shape, to be of unit mass and to have the same sedimentation rate.

Aerosol
An atmosphere containing particles which remain airborne for a reasonable length of time

Black Smoke
Non-reflective (dark) particulate matter, associated with the smoke stain measurement method

Brownian motion
Constant small movement of suspended bodies due to bombardment by surrounding molecules

Coarse particle mode
A part of the size spectrum of airborne particles, greater than about 2 µm diameter, in which particles have arisen mostly from disintegration of bulk solid and liquid materials

Coarse Particles
Particles within the coarse particle mode. In this report, often refers to particles in the 2.5-10 µm size fraction.

Count Median Diameter (CMD)
Term used to characterise the size distribution of particles in an aerosol. 50% of particles are of diameter less than the count median diameter

Fine Particles
Particles smaller than about 2 µm diameter which arise mainly from condensation of hot vapours and chemically-driven gas to particle conversion processes. In this Report, often refers to the $PM_{2.5}$ fraction.

Geometric standard deviation
Term used to describe size distribution of particles conforming to a log-normal function

Hygroscopic growth
Growth of particles due to uptake of water from the atmosphere

Inhalable particles
Particles which may be breathed in. "Inhalability" is the orientation-averaged aspiration efficiency of the human head. (Also termed inspirable)

Mass concentration
The concentration of particles in air expressed as mass per unit volume

MMAD
Mass Median Aerodynamic Diameter. Term used to characterise the distribution of sizes of particles in an aerosol. MMAD is the size where half the mass of the cloud is contained in particles of aerodynamic diameter smaller than the stated aerodynamic diameter and half in larger particles

MMD
Mass Median Diameter. As for MMAD except that the diameter considered is the actual diameter rather than the aerodynamic diameter

Monodisperse
Term used to describe an aerosol containing particles of only one size

Nucleation mode
A part of the size spectrum of airborne particles, below about 100 nm diameter, in which particles arise mostly from fresh emissions from combustion processes, and gas to particle conversion

Number concentration
The concentration of particles in air expressed as number of particles per unit volume

Number median diameter (NMD)
Term used to characterise the distribution of sizes of particles in an aerosol. The NMD is that size where half the total number of particles are of diameter greater than NMD and half of diameter less than NMD (same as count median diameter)

Photochemical smog
Smog caused by the formation of particles due to a chemical reactions driven by sunlight

$PM_{2.5}$
Particulate matter less than 2.5 µm aerodynamic diameter (or, more strictly, particles which pass through a size selective inlet with a 50% efficiency cut-off at 2.5 µm aerodynamic diameter)

PM_{10}
Particulate matter less than 10 µm aerodynamic diameter (or, more strictly, particles which pass through a size selective inlet with a 50% efficiency cut-off at 10 µm aerodynamic diameter)

PM_{15}
As PM_{10}, but with 15 µm as the cut-off

ppb
Parts per billion, 1 part by volume in 10^9

ppm
Parts per million, 1 part by volume in 10^6

Relative humidity
Actual vapour pressure/saturated vapour pressure expressed as a percentage. A measure of the degree of saturation of the air with water vapour

Respirable Particles
Particles which can penetrate to the unciliated regions of the deep lung.

Smoke
Particulate matter, <15µm, derived from the incomplete combustion of fuels.

Suspended Particulate Matter (SPM)
A general term embracing all airborne particles.

TEOM
Tapered Element Oscillating Microbalance, an instrument for the continuous measurement of suspended particulate matter in air

Thoracic particle mass
Describes that fraction of the particles which penetrates beyond the nasopharynx and larynx

Total Suspended Particulate (TSP)
A term describing the gravimetrically determined mass loading of airborne particles, most commonly associated with use of the US high volume air sampler in which particles are collected on a filter for weighing

Troposphere
Layer of the atmosphere extending upwards from the earth's surface for about 10 km

TSP
Total suspended particulate

Ultrafine particles
Particles of less than 50 nm diameter (some workers use 100 nm)

Terms of Reference and Membership

TERMS OF REFERENCE

1. The UK Review Group on Urban Air Quality is a working group of experts established by the Department of the Environment to review current knowledge on urban air quality and to make recommendations to the Secretary of State for the Environment.

2. The initial objective of the Group is to prepare a review of urban air quality and how it is assessed in the United Kingdom especially in relation to public exposure, and how this information is passed on to the public. To this end the Group will consider:

   i) the pollutants measured,

   ii) the extent of monitoring networks,

   iii) the consistency of data,

   iv) the types and location of monitoring equipment and

   v) any other relevant material.

3. The longer term objectives of the Group will be to:

   i) perform a rolling review of the subject in the light of scientific and technological developments,

   ii) consider, in the light of national and international guidelines and advice, the need to add or subtract sites from the networks and the need for additional networks for different pollutants and

   iii) to consider arrangements for the public availability of data.

4. The Group will identify areas of uncertainty and recommend where further research is needed.

5. The Group will make recommendations for changes to relevant monitoring networks and public information systems.

6. The Group will act as an informal forum for the discussion of research plans and results.

7. The Group will act as a point of liaison with relevant international bodies.

MEMBERSHIP

Professor R M Harrison (Chairman)
Institute of Public and Environmental Health
School of Chemistry
The University of Birmingham
Edgbaston
Birmingham B15 2TT

Professor P Brimblecombe
School of Environmental Sciences
University of East Anglia
Norwich NR4 7TJ

Dr R G Derwent
Meteorological Office, Room 156
Met O (APR)
London Road
Bracknell
Berkshire RG12 2SY

Dr G J Dollard
AEA Technology
National Environmental Technology Centre
Culham
Abingdon
Oxfordshire OX14 3DB

Dr S Eggleston
AEA Technology
National Environmental Technology Centre
Culham
Abingdon
Oxfordshire OX14 3DB

Professor R S Hamilton
Urban Pollution Research Centre
Middlesex University
Bounds Green Road
London N11 2NQ

Mr A J Hickman
Transport Research Laboratory
Old Wokingham Road
Crowthorne
Berkshire RG11 6AU

Dr C Holman
Brook Cottage
Camp Lane
Elberton
Bristol BS9 2AU

Dr D P H Laxen
Air Quality Consultants
12 St Oswalds Road
Bristol BS6 7HT

Mr S Moorcroft
TBV Science
The Lansdowne Building
Lansdowne Road
Croydon CR0 2BX

Observers:

Dr S Coster
Ms L Edwards
Air Quality Division
Department of the Environment, Room B355
Romney House, 43 Marsham Street
London SW1P 3PY

Dr R L Maynard
Department of Health
Hannibal House
Elephant and Castle
London SE1 6TE

Secretariat:

A R Deacon
University of Birmingham
Institute of Public and Environmental Health
School of Chemistry
Edgbaston
Birmingham B15 2TT